Measuring the Quality of Care

RECENT ADVANCES IN NURSING

Already published

Current Issues in Nursing
Edited by Lisbeth Hockey

Care of the Aging
Edited by Laurel Archer Copp

Cancer Nursing
Edited by Margaret C. Cahoon

Nursing Education
Edited by Margaret Steed Henderson

Primary Care Nursing
Edited by Lisbeth Hockey

Patient Teaching
Edited by Jenifer Wilson-Barnett

Communication
Edited by Ann Faulkner

Perinatal Nursing
Edited by Peggy-Anne Field

Maternal and Infant Health Care
Edited by Mary J. Houston

Forthcoming Titles in the Series

Perspectives on Pain
Edited by Laurel Archer Copp

Psychiatric Nursing
Edited by A.T. Altschul

Long-Term Care
Edited by Kathleen King

Research Methodology
Edited by Margaret C. Cahoon

Nursing Practice
Edited by Alison J. Tierney

RECENT ADVANCES IN NURSING   10

Measuring the quality of care

Edited by

Lucy D. Willis RN BS MA EdD
Professor of Nursing
College of Nursing
University of Saskatchewan
Saskatoon, Saskatchewan
Canada

Marjorie E. Linwood RN BSN
Associate Professor of Nursing
College of Nursing
University of Saskatchewan
Saskatoon, Saskatchewan
Canada

CHURCHILL LIVINGSTONE
EDINBURGH LONDON MELBOURNE AND NEW YORK 1984

CHURCHILL LIVINGSTONE
Medical Division of Longman Group UK Limited

Distributed in the United States of America by
Churchill Livingstone Inc., 1560 Broadway, New York,
N.Y. 10036, and by associated companies, branches
and representatives throughout the world.

First published 1984
  Reprinted 1986

ISBN 0-443-02920-2
ISSN 0144-6592

British Library Cataloguing in Publication Data
Measuring the quality of care.
  (Recent advances in nursing, ISSN 0144-6592; 10)
  1. Nursing audit
  I. Willis, Lucy D. II. Linwood, Marjorie E. III. Series
  610.73 RT85.5

Library of Congress Cataloging in Publication Data
Measuring the quality of care.
  (Recent advances in nursing; 10)
  Includes index.
  1. Nursing — Quality control. I. Willis, Lucy D.
II. Linwood, Marjorie E. III. Series. [DNLM: 1.
Nursing care — Standards. 2. Outcome and process
assessment (Health care) 3. Quality assurance, Health
care. W1 RE105VN v. 10/WY 16 M484]
RT85.5.M435 1984 362.1'73 84-1727

Produced by Longman Singapore Publishers Pte Ltd
Printed in Singapore

Preface

For more than two decades nurses have been sealing the fate of the admonition 'You can't measure quality of care.' This volume is an attempt to bring together information of the development that has taken place in the field in a form that makes it both enlightening and useful at management and individual caregiving levels of nursing.

The first two chapters are introductory. Van Maanen discusses measurement and evaluation of the quality of nursing care as professional function and responsibility with the complementary component, accountability. Care and caring are discussed using a series of models to demonstrate changes in these concepts over time. The consequences of these changes for both nursing and consumers are explored. The subject of standards and criteria is introduced and illustrated, duodirectionally, employing consumer needs and the state of professional expertise.

Rosso has taken a different stance — the state of clinical research and its present and potential role in improving direct patient care. She speaks from a background as research co-ordinator of an advanced acute care centre and, as consultant to a provincial association, advisor to nurses in rural and city hospitals. She illustrates the application of basic research ideas in the practice setting and discusses what to expect when a research committee is established with change as its goal. Establishing standards and auditing are ways of improving care using the rigorous methods and organization of research.

In Part Two, Chapters 3, 4, and 5 are illustrative of major developments in the quality assurance movement. Primary movers were a number of academics and nursing practice leaders in the United States. Lang and Clinton, who are of this company, discuss the concept of quality assurance and the use of a model to explore the fundamental processes involved. The history of the movement is briefly outlined and documented. Specific quality assurance measurement tools are introduced. The chapter goes on to discuss the impact of these developments on specific nursing care situations and the public policy implications.

Jacquerye and Jenkins write from different experiences of establishing quality assurance programs in Belgium and Australia respectively. Knowledge and understanding of the processes and of the setting for implementation are crucial. Jacquerye explores both anticipated gains and management concerns implicit in several prototype institutional programs. Jenkins outlines the procedures and processes that were undertaken by the Royal Australian Nursing Federation in a national unified approach to instituting a quality assurance program. Involvement, motivation, and acceptance were fostered by a judicious introduction of ideas. Standards and criteria were at the proposal stage at time of writing.

In Part Three, Giovannetti presents work that has been done on the allocation of nursing staff using patient classification tools. Hilton reports on studies of sleep, noise, and diabetic monitoring using mechanical instrumentation to evaluate quality of nursing care. Henney describes a real-time computerized nursing system designed to facilitate optimum delivery of nursing care. These three chapters provide a glimpse of some of the specific developments that are closing the gap between knowledge base and practice.

The annotated bibliography in Part Four has been sectioned according to probable approaches to quality improvement and usage. References cited in individual chapters have not been repeated so readers should consult these also in their search for pertinent literature.

Saskatoon, 1984   L.D.W.
                                                           M.E.L.

Contributors

Jacqueline F. Clinton RN PhD FAAN
Associate Professor and Center Scientist, School of Nursing, University of Wisconsin-Milwaukee, Wisconsin, USA

Phyllis Giovannetti RN BN ScD
Associate Professor, School of Nursing, University of Alberta, Edmonton, Canada

Christine R. Henney RGN SCM
Research Nursing Officer and Honorary Research Fellow, Department of Therapeutics and Pharmacology, University of Dundee, Ninewells Hospital and Medical School, Dundee, UK

B. Ann Hilton RN BSN MScN
Assistant Professor of Nursing, School of Nursing, University of British Columbia, Vancouver, Canada

Agnès Jacquerye
(Nurse, Master's Degree Hospital Sciences; Degree in Statistic Epidemiologic and Operational Methods in Medicine and Public Health) Chief of Inservice Education Service, Department of Nursing, Cliniques Universitaires de Bruxelles, Hôpital Erasme, Bruxelles, Belgium

Enid R. Jenkins RN DipNseEd BASc
Federal Office Professional Development Officer, Royal Australian Nursing Federation, Melbourne, Australia

Norma M. Lang RN PhD FAAN
Dean and Professor, School of Nursing, University of Wisconsin-Milwaukee, Wisconsin, USA

Margaret J. Rosso RN BN MSc (Nursing)
Nursing Consultant, Saskatchewan Registered Nurses Association, Regina, Saskatchewan, Canada

Hanneke M. Th. van Maanen RN MA
Doctorate student in Nursing Sciences, University of California, San Francisco, and Master's student in Public Health, University of California, Berkeley, USA. Formerly Head of Department of Nursing Studies, National Ziekenhuis Institut in Utrecht, The Netherlands.

Contents

PART ONE   Evaluation of Nursing Care

1. **Evaluation of nursing care: a multinational perspective**   3
    Hanneke M. Th. van Maanen

2. **Knowledge for practice: the state of clinical research**   43
    Margaret J. Rosso

PART TWO   Quality Assurance Movement

3. **Quality assurance — the idea and its development in the United States**   69
    Norma M. Lang and Jacqueline F. Clinton

4. **Quality assurance in nursing: the Australian experience**   89
    Enid R. Jenkins

5. **Choosing an appropriate method of quality assurance**   107
    Agnès Jacquerye

PART THREE   Research to Change Outcomes of Care

6. **Staffing methods — implications for quality**   123
    Phyllis Giovannetti

7. **Answering the question with electronic instrumentation**   151
    B. Ann Hilton

8. **The use of computers for improvement and measurement of nursing care**   174
    Christine R. Henney

PART FOUR   Bibliography   191

Index   199

PART ONE

Evaluation of Nursing Care

1

Hanneke M. Th. van Maanen

Evaluation of nursing care: quality of nursing evaluated within the context of health care and examined from a multinational perspective

Their care is our concern
Logo WENR

The development of nursing as a profession with an emerging body of knowledge has resulted in a growing interest for the improvement of quality nursing care among nurses from all nations. The interest is founded in the commitment to people they serve and shown in programs which aim toward the maintenance and promotion of health and well-being.

Although the beliefs are rooted in the same philosophy, the means to accomplish these goals may differ from country to country. This chapter presents an overview of the evaluation of nursing care from a multinational perspective. The content is limited however to the author's exposure to nursing systems and professional performance in primarily industrialized coountries, among them Belgium, Sweden, The Netherlands, United Kingdom and the United States. This approach is restrictive since nursing in affluent societies has particularly developed in specialized health care institutions (Smit, 1981). In the majority of Western countries the focus of nursing is heavily skewed towards the acute care delivery system. However, nursing has a much broader scope. Nurses in their capacity as care specialists also play a crucial role in the delivery of community health services. Community nursing, an integral part of primary health care, has developed in a number of industrialized countries, frequently as an extended service offered by secondary or tertiary health care facilities. Sometimes it started as a community service based on private initiative or facilitated within the structure of socialized health care. In many countries health care seems to be regarded as identical to sick care and the delivery of services centers in and around acute hospitals (Bandman & Bandman, 1981). This is in contrast to the Third world countries where community health services tended to develop in congruence with the emerging needs of

3

the population served, before sophisticated health care institutions could be established (Annel, 1982).

If an entire book is devoted to quality of nursing care, the question about the focus of its contents is a legitimate one. The subject can either be discussed within a framework of health care as a total entity of which nursing is one of the major cornerstones or it can be discussed within a frame that is restricted to those practice settings where care has already been monitored for quality as in a number of hospitals, Scheme I, (Fig. 1.1). Although all health services should be reviewed for quality, in practice most quality assurance programs are found in hospitals, a convenient unit of analysis with relatively easy access to data (Lacronique, 1981).

Since the methods of quality monitoring will be discussed in the second part of this volume, my preference is to discuss the quality of nursing care and its evaluation in a broader context. The rationale for this choice is that any method or system of quality evaluation is based on some principles and conditions. If nurses agree on those underlying principles, the methods will flow from there and can be developed according to the needs and characteristics of a special care facility, be it the community or a health care institution.

Another reason is that in a time of increasing financial restrictions there will be limited funding for the development of sophisticated

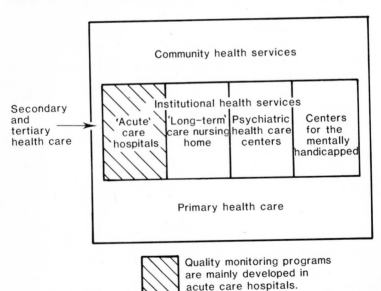

Fig. 1.1   Scheme I Health care delivery system (simplified)

Primary health care is essential health care based on practical, scientifically sound and socially acceptable methods and technology made universally accessible to individuals and families in the community through their full participation and at the cost that the community and the country can afford to maintain at every stage of their development in the spirit of self-reliance and self-determination. Declaration of Alma-Ata (Hogarth, 1978).

Secondary care★★ comprises the care provided through specialized services on referral from primary care services ...

Secondary medical care requires attention of a special nature, usually more sophisticated and complicated than can be handled by the general practitioner.

Tertiary care★★ includes highly specialized services and eventually the super-specialties ...

Tertiary medical care requires highly specialized attention, which can usually only be provided in centres specially designed for the purpose, and by physicians trained in the area of specialization. (WHO Expert Committee on Planning of Medical Education, 1974, p. 182 in Hogarth, 1978).

Community *health* care refers to the health status of the members of the community, to the problems affecting their health, and to the totality of health care provided for the community. (WHO Community Health Nursing, 1974, p. 8).

Community *nursing* includes family health nursing but is also concerned with identifying the community's broad health needs and involving the community in development projects related to health and welfare. It helps communities to identify their own problems, to find solutions, and to take such action as they can before calling on outside assistance. (WHO Community Health Nursing, 1974, p. 11).

Institutional health services

Acute care hospitals — Teaching hospitals (university hospitals), regional and community hospitals;

Long-term care hospitals — Hospitals for non-acute care, nursing homes, rehabilitation centers, etc;

Psychiatric hospitals and centers — A variety of institutional services covering acute-, medium- and long-term-care, daycare centers, sheltered living, etc;

Centers for the mentally handicapped — Institutions (larger) and homes (smaller) for the mentally handicapped, sheltered living.

These four broad categories of health care institutions can be complemented with other specialized intramural services.

★★ The definitions of secondary and tertiary care are strongly focused on medical performance. It is assumed that for that reason the concept of health is eliminated from both definitions.

monitoring systems. Although the development of evaluation systems should not be restricted by the available budget, the reality is that not all nurse-managers have control of the budget spent for nursing services. The lack of power to regulate nursing practice is a serious problem in modern nursing (Scott Wright, 1983). Nurses will need all their creativity and imagination to develop evaluation systems that are effective, efficient, and available at low cost. No first class nursing care can be delivered under third class conditions!

With regards to the rapidly rising costs, health workers, including nurses, should investigate why their services have become so over-utilized that the balance between supply and demand has grown out of hand, especially in countries with a well developed system of social services (Wiener et al, 1982). The recession has served as a reality shock, needed to discover limitations in what society can afford. The financial situation has called a halt to the supply, even resulted in severe cut-backs in services. A decrease in demand is forced upon the population, but is not realized overnight. It is a painful process of adjustment and requires a drastic change in mentality throughout which the rights to health care of vulnerable groups in society should be safeguarded. Resources have to be re-evaluated against the identified needs for health care. With the trend to strengthen primary health care services, the responsibility for the maintenance of health is returned to the consumer, who needs support and instruction to adopt this long forgotten role.

CHANGES IN THE CONCEPT OF CARE AND CARING

The interest for quality control and improvement is related to advances in health care that have taken place over the last three decades. Although there has always been a professional commitment to provide quality services, the development of some philosophy on care and caring and the structuring of those thoughts in a framework for measurement and evaluation is of more recent origin.

In the past, health care has been used synonymously with medical care delivered in hospitals, that over the years changed from caring institutions to centers for diagnosis and treatment. Think of the number of hospital gardens that were replaced by high rise intensive care units! The advances in medical technology raised the ethical question about means and purpose; did technology serve the patient or was the patient subject to technology? (Eldar, 1982).

A growing awareness among consumers and health professionals about the challenges and risks of a technocratic health care system has resulted in evaluation of present medical services in a broader context of health care requirements. The major question is what kind of health services do people need and how are these needs met by society and the responsible health professions? A more recent issue is, all people have the right to health,* but at what level, of what quality and at what costs? (World Health Organization/UN-

---

* It should be noted that the Alma-Ata Declaration mentions health instead of health care services. For affluent societies this is an implicit consequence, for other nations a goal that cannot be accomplished without support.

ICEF, 1978, Alma-Ata Declaration, p. 3).

In one part of the world people's lives can be maintained and prolonged with the implantation of artificial organs and the use of life-saving equipment, in other parts of the world, the maintenance of life is at risk simply because of the lack of food and shelter. Health care is a luxury in places without access to the most minimal resources (UN/WAA, 1982). The crucial question of the 1980s will be how the available resources are to be divided. This is a political decision and a matter of determining priorities: care for all at the cost of a few or care for a few at the cost of many? (Peccei, 1983; van der Velden, 1983).

The availability and the quality of health care are determined by the values and expectations of consumers and among them the health professions. The consumer expects value for his money and counts on the existence of services when he needs them (Wiener et al, 1980). In a number of countries consumers are becoming more critical of the health care provided. They have united themselves in consumer's organizations and claim rights as active participants in the planning and evaluation of health services. The Grey Panthers, a nationwide organization of aging persons in the USA is an example.

The role of the health professionals, although potential consumers, is slightly different. They have similar expectations but their responsibility and accountability reaches beyond the point of consumption. The accountability is accomplished by regulatory mechanisms of education and licensure, legislation, the adoption of a code of ethics and the control of professional performance (e.g. peer review). These regulatory mechanisms developed from within the profession and accredited by legislative authorities guarantee a defined level of quality services to the public.

The changes in the concept of care and caring are illustrated with the following models. The advantage of a model is that it provides a framework for reflection and study, the disadvantage is that it presents the context in an absolute manner, whereas in reality developments overlap each other. In the clinical area, elements of each of the discussed models may be observed at the same time.

Model I

Most care is still provided in health care institutions with a strong focus on medical technology, reflected in the emphasis on diagnosis and treatment of somatic disorders. Although the goal of treatment is health, it can be questioned whether elimination of problems in body functioning is identical to the restoration of health. The frequency of re-admission is just one indicator to evaluate the effectiveness of the services provided.

The organizational context in which the care is delivered tends to be ritualized, routinized and fragmented (van Maanen, 1979). Each discipline cares for a segment and cares less for the person as a whole. Goal-directed leadership and an hierarchical structure are the means to link the fragmented services. In order to limit costs, programs for screening and treatment have been condensed to a minimum number of hospital days per case. Health workers may have difficulty coping with the increasing workload, but what about the poor patient? According to Hoenson (1972) the patient is exploited by making working days of 18 hours or more for the sake of whom? His meals are interrupted for tests and he is awakened at night to have his blood pressure taken and so on.

Model II

The limitations of the discipline-oriented medical model concerned not only groups of organized consumers and social scientists, but also health professionals who became aware that sophisticated medical services did not raise the level of health in the population. More interest developed in the patient as a holistic person, whose health problems might affect part of his functioning, but should not restrict functions that could be maintained. The more or less passive consumer of care was challenged to become an active participant.

The patient/client oriented approach of care to which each health worker was expected to contribute, required an adjustment in the education of the health professions. Social sciences were included in curricula and more attention was paid to the mastery of communication skills. Technique oriented care was complemented by more appreciation for the psychosocial needs of a person. Since no health worker could cover the full range of services any longer, people started to refer to each other, at first from within their own discipline, and later among disciplines. Multidisciplinary relationships evolved, characterized by delegation of responsibility and stronger social leadership.

Model III

The change from holistic to emancipatory care is based on the philosophy that each individual, whether patient or client is entitled to health and a state of well-being. This is accomplished through an open relationship between consumer and health professional(s), the exchange of information and the right to make decisions, even if such decisions should conflict with the norms and values of the responsible care provider. In accordance with the wishes of the patient/client, the family or significant other persons are included in this process.

There is a horizontal relationship between the care provider and the care recipient. The role of the health professional is a supportive one in an effort to stimulate the abilities for self-care and independence. The co-ordination of services requires leadership skills in a change-agent capacity with emphasis on interpersonal relations and the re-balancing of preventive, curative and restorative care.

Table 1.1 Change in the concept of care and caring (Boekholt 1982; translated by van Maanen)

	Model I	Model II	Model III
Concept	treatment	counselling and guidance	self-care and supportive care
Scope	medical-somatic	holistic	emancipatory
Patient/client	medical case; dependent	individuals; cooperative	individual in social context; partner in decision making
Goal	health	health and well-being	health, well-being and awakening of consciousness, resocialization
Professional role	medical-technical	patient/client oriented; comprehensive care	patient/client and system oriented
Focus education	discipline and technique oriented	technique oriented; communication skills	technique oriented; communication skills and understanding of socio-economic relationships and community engagement
Organizational context	hierarchical; segmented; discipline oriented; routinized	open communication; consultation; multidisciplinary relationship	open communication; participation of patient/client and family in consultation; partnership
Leadership	goal-directed	delegation; social leadership	social leadership and change agent role

The approach is outreaching in the sense that the delivery of health services is not restricted to health professionals, but encompasses the community and its support systems. Examples are the integration of a handicapped person in the community and the support services made available to frail elderly.

CHANGES IN NURSING CONCEPTS OF CARE AND CARING

After having examined the changes of care and caring in a primarily institutional care context, this part will focus more closely on the consequences of these changes for nursing, expanding the scope to the community as well (Models A, B, C).

Nursing, a humanistic science, dates back as far as ancient Greece, a society that did not make a distinction between nursing and medicine. Plato, the founder of scientific medicine, identified four areas of care: pharmacology, surgery, assistance of care and knowledge of disease and diagnosis. Minute descriptions of assistance of care characterize modern nursing (Lanara, 1981; 1982). The curing care was holistic, comprehensive and included the patient and his environment. The era between ancient Greece and the 19th century offers limited significant information on nursing other than from a charitable perspective. It was not until 1850 that, under the inspiring leadership of Florence Nightingale, nursing could develop the characteristics of a discipline. The Victorian Age determined the direction in which nursing, a women's occupation, would progress.

Model A

Until the First World War, care of the sick was primarily delivered at home. With the development of medicine, home-based care was gradually taken over by hospitals with a focus on diagnosis and treatment. Nursing followed the medical trends and adapted to the organizational structure of the institutions. In order to keep up with the advances in medical sciences, nurses acquired new skills to provide patient care that would complement a more technological approach to medical practice. The humanistic aspects of nursing were overruled by the challenges of science and the rewards offered upon involvement in problem analysis and case finding with the help of monitors, scanning and clinical chemistry.

The care for the sick diminished to cure of the sickness, signs and symptoms were classified in terms of somatic medical terminology. The recipient of care was exposed to segmented services delivered by a large number of different health professionals. Nurses who were expected to safeguard the co-ordination and continuity of care picked up the loose ends where other professions let go. The rigid structure of the hospitals, a reflection of the level of management and the influence of management consultants, resulted in the introduction of industrial methods of quality control and efficiency. The technics may have been sophisticated for industry, but could not be applied to a human product, namely the patient. However, the side-effect of goal-directed management was task allocation and time control. Nurses became victims of their willingness to expand their role beyond nursing: the care was ritualized and routinized, the only alternative to survive the system.

Although similar trends became visible in community nurrsing e.g., in the centralization of management, the introduction of patients' records and the segmentation of nursing expertise, the nature of nursing still allowed rather independent practice. It should be noted that the delivery of community nursing services among countries shows more variation than nursing care delivered in hospitals.

Model B

The penetration of the social sciences in the fortress of health care and in particular the hospitals, threw a new light on the quality of caring being practiced. Nurses expressed concern about the lack of time devoted to contact with the patients/clients and searched for new approaches in the delivery of nursing services (Baumgart, 1982). Team nursing seemed to be the answer. A small group of nurses under the leadership of the most senior nurse was assigned to a group of patients. This model could have improved and personalized the care if the team members would have been of equal strength, that is professional nurses, and able to function in a peer relationship. This was not the case since most institutions are staffed by at least three categories of nursing personnel with a variety in educational background from a professional level to training on the job. Since all these workers are addressed as nurses, it is not surprising that the public as well as other healthworkers become confused in recognizing the characteristics of professional nursing practice. Who is who? (Henderson, 1978; Hall, 1979; Hall & van Maanen, 1981).

The employment of various categories of nursing personnel and the shortage of registered nurses, resulted in the composition of teams at the best headed by a registered nurse, but frequently staffed by learners (students) and less qualified personnel. The meaning of

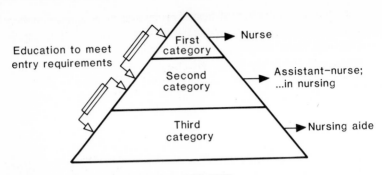

Fig. 1.2   Nursing personnel subsystem (Hall, 1979)

the word team is in this situation closer related to the structure of organization than to the content of nursing practice.

However team nursing was an improvement in the sense that it facilitated a climate for more personal care and a closer communication with the patient. Although nursing services were still delivered within the context of the medical model, nurses gained a stronger appreciation for the emerging body of knowledge within their own discipline. It was the phase of the development of assessment instruments and nursing records.

Team conferences changed their scope from task allocation to the planning and evaluation of patient/client care. Nurses gained expertise in working together in a collegial manner. The approval and support of its leadership was of crucial importance for the success or failure of team experiments. The process oriented approach to care delivery required flexible management operationalized in social leadership.

Team nursing is nothing new in community nursing where nurses function in more interdependent roles. Patient conferences, community projects and weekend coverage are just a few examples.

Model C

Nurses were not satisfied with the overall outcome of team nursing. In some instances it seemed to be identical to task oriented nursing in a more person focused environment. The group to group: nurses versus patients approach did not have the characteristics of professional practice, the main problem being that the care was still fragmented.

New models emerged, the most recent concept being primary nursing, described by Hegyvary (1982, p. 3) as 'both a philosophy of

care and an organizational design. It is a view of nursing as professional, patient-centered practice.'

The characteristics are:

> *accountability:* the primary nurse carries around-the-clock re-
> sponsibility for the nursing care delivered to the patient. In
> community nursing the nurse would be answerable during 24
> hours a day;
> *autonomy:* the primary nurse has the authority to make decisions
> with regards to nursing care provided to the patients;
> *coordination:* the primary nurse acts as a 'trait d'union' (link)
> among caregivers, who are involved in the care for the patient/
> client. In an institution the co-ordination centers around medic-
> al services, whereas in the community the nurse may facilitate
> co-ordination of health and social services, particularly in cases
> of long-term illness and rehabilitation;
> *comprehensiveness:* the primary nurse facilitates individualized
> patient care throughout a specific time period (Hegyvary,
> 1982).

There are several modelities of primary nursing. However the
main characteristic of the concept is the personalized relationship
between the primary nurse and the recipient of care, reflected in
nursing care that is continuous, co-ordinated, comprehensive, individualized and patient/client centered (Marram et al, 1979). It involves the family or significant others at the recipient's discretion.
The transition from team nursing to primary nursing is in essence
the process of development from vocational to professional nursing
with all the changes in values and beliefs of a young profession
offering its services to a society that sooner or later will accredit its
advancements.

The development of primary nursing does not only have consequences for the quality of care given to the patient/client, but also for
the community because the needs of an individual are evaluated in a
broader context of professional services made available to people in
health and illness. Comprehensive or all-round, in-depth care, facilitates early identification of needs which may be observed in one
person, but be of relevance for the health condition of others. An
example is the care for the drug addict.

Holistic nursing focuses on the health and well-being of people. It
recognizes the person's ability for self-care and decision making and
assists the individual in coping with responses to illness and disease.
This is done by complementing deficits in self-care by supportive

care, by strengthening the functional abilities and by teaching. The primary nurse and the patient/client learn with and from each other. The skills required at the management level are defined as social leadership in a change agent capacity (MacPhail, 1972; van Eindhoven, 1979; Koene et al, 1980).

Primary nursing is research based. It embodies scientific concepts emerging from nursing practice that can be tested for their relevance in nursing. It uses new knowledge and contributes to the develop-

Table 1.2   Changes in the nursing concept of care and caring (van Maanen, 1983)

Model A	Model B	Model C
Concept task-oriented	team nursing	primary nursing

Scope
  medical/nursing
  somatic
	nursing/medical
	biopsychosocial
		nursing/medical
		biopsycho social

Patient/client
  medical and
  nursing care;
  dependence
	nursing and medical care to
	individual; cooperation
		nursing holistic; focussing
		on recipient's independence;
		interdependent relationship;
		care provider and recipient

Goal illness to health	illness to health and well-being	health and well-being as focus for care and cure

Professional role
  nursing —
  technical
	patient/client oriented;
	person directed technical
	care
		comprehensive professional
		nursing care complementing
		person's ability for self-care
		and decision making

Focus education
  job training
  discipline and
  technique oriented
  with emphasis on
  medical approach
  of cure and care
	discipline oriented;
	communication skills
	complement technical
	approach; nursing awareness
	of care and cure
		profession oriented; nursing,
		a health-oriented discipline
		meeting the needs of the
		people it serves; the needs of
		individuals related to the
		community/society

Organizational context
  hierarchical;
  segmented;
  task-oriented:
  ritualized,
  routinized
	open communication; team
	conferences; multidiciplinary
	relationship
		open dialogue between nurse
		and patient/client and family;
		relationship based on
		partnership; practice
		research based

Leadership
  goal-directed
  authoritarian
	referral; social leadership	delegation and consultation;
		social leadership in change
		agent capacity

ment of knowledge by formulating hypotheses.

New knowledge should be tested rigorously for its validity, before it can be applied to practice. However this phase of research has not always been valued enough. *Models* and *frameworks* for nursing have been adopted in the process of construction, evaluation and refinement. Some of them have been labelled *theories* by nurses who were eager to learn more about the nature of nursing, but not familiar with the requirements of theory development. Although the models contributed to the critical evaluation of nursing practice and education more study was required before their relevance for nursing could be determined from a scientific perspective. This is done retrospectively.

Roy's adaptation model (1976) and Rogers' model of unitary man (1970) are two examples of frameworks that justify thorough testing and could contribute insight in nursing as an emerging science.

In the first part of this chapter, health care and nursing have been evaluated within the context of health care needs and professional (r)evolution. The second part will focus on the principles of quality appraisal and evaluation as they apply to nursing.

QUALITY

Quality is a concept that is frequently used in societies with a high living standard, countries that can afford to expand their investments in the improvement of human and material resources. It is an abstraction defining the margin between desirability and reality, the ideal and the reality. It is the defined versus the observed quality, the latter operationalized in terms of measurable quantities (van Maanen, 1979).

The concept *quality* is part of every day jargon and often used in a rather casual manner. Sometimes it is described in certain characteristics: e.g., an orange ready for consumption is ripe, juicy, large and sweet; sometimes it is subject to value judgments: a bad orange. The characteristics can be observed and quantified, an important condition to do empirical research. Value judgments don't have these properties because they label phenomena as they are perceived by people.

It is not simple to develop a framework within which quality can be evaluated. People have their own opinions based on norms, values and expectations. In order to live and work together people have to agree on basic rules with regard to what is acceptable, that is meeting the norms of society, and what is not. It depends on the kind of society, its culture and tradition whether the boundaries of social acceptance are flexible or not.

Standards of quality

The evaluation of quality of care in itself is an abstraction and needs to be operationalized for application in practice. The tools for evaluation are standards that serve as points of reference in the process. Standards are developed by people and are the result of naming and discussing the properties of a concept or object. Decisions have to be made with regard to the characteristics of quality expressed in concrete terminology. Although people may give different weight and value to the concept or object, they have to come to some agreement in terms of standards. The final judgment on standards should be approved by all involved in the evaluation. It is based upon knowledge, experience, interest and priorities. The development of standards is a dynamic process that requires continuing updating and adjustment.

Evaluation can take place in relation to people and products. It is a complex procedure to evaluate quality in relation to actions and behaviour of people since each individual is unique and modelled by a variety of internal and external factors that cannot easily be classified in quantities and qualities. Besides, what would be the purpose of it? Does such evaluation serve the person, the society or is it an exponent of personal norms and values?

Example 1
Standards can be used to measure student's progress in school to see whether the student is accomplishing his learning goals or whether the learner's competencies are congruent with the instructor's teaching goals. The evaluation of student's progress serves as reference point for the planning of courses and individual careers.

Example 2
If an early morning subway is overcrowded by people whose motionless faces and bodies ignore the presence of others and suddenly a man stands up, shouting while waving his arms in agony 'Can nobody say good morning to anyone? Who are you? Are you people alive?', then the question arises whose behaviour is normal and acceptable and according to which norms? The evaluation of this incident tells us only about modes of interaction among people but on such a limited scale

that no inferences can be drawn from them.

The evaluation of quality of a product is less difficult to determine. The characteristics are measure against the standards, defined before the production took place. Some properties are vital for its function, others are important, but less relevant for its use.

Example 3

A car with three instead of four wheels cannot be driven safely. However, its main function, driving, is not restricted because it has two instead of more coats of paint before it is delivered for sale.

Standards of quality of care

There is no consensus on the meaning or use of measuring standards of nursing care. Do nurses refer to the highest attainable level of professional practice, an abstraction, or to the quality of care that is delivered? Crow (1981) observed that 'standards of care' is used synonymously with both quality of care and effectiveness of care. This is confusing particularly for research purposes when concepts such as standards, quality and effectiveness should be clearly distinguished. If concepts are used synonymously or interchangeably, the validity of the research is at risk. This is certainly the case when evaluation relies on the value judgment of surveyors rather than researchers. Although standards of evaluation should be flexible, some kind of measurement instrument is needed to determine the levels of care that are accomplished. A validated scoring system should measure the outcome (McAlary, 1981; Padilla & Grant, 1982).

The concept *standard* is used as a measure of conformity, a basis for comparison and as a level of performance in professional practice. A working definition that incorporates all the characteristics of quantitative and qualitative measure is:

> Standards of care is some measure or measures by which nursing care can be judged or compared and where the measures used are agreed upon by common consent (Crow, 1981).

A model for evaluation

The model used most frequently is based on three interrelated domains: structure, process and outcome (Donabedian, 1976).

Structure: evaluation of the organization of the institution delivering care; the conditions under which care is provided and its impact on the quality in e.g., manpower and material resources like buildings, budget, equipment. The use of resources is interlinked by coordination, continuity and accessibility.

Process: evaluation of the performance of nurses, physicians and other health professionals in the management of patients and clients. The professional activities are evaluated while the care is concurrent, in process, using predictable, prospective criteria.

Outcome: evaluation of the end results, observable changes in the health status of the patient/client. The outcome is evaluated by using a retrospective approach, based on individual criteria (Mayers et al, 1977; Lorensen, 1979).

Berg (1974) uses the term content of care rather than process. In focusing on professional performance, the question is raised whether nursing is practiced according to the standards of the profession.

Structure	→	Process	→	Outcome
		or		
Input	→	Throughput	→	Output
		or		
what can be offered?	→	how are resources used?	→	what end-result is expected or accomplished

Standards and criteria

Each domain can be divided into subdomains. Hagen (1976) has expressed concern about the interchangeable use of two important concepts: standards and criteria. She defines:

standards as a cluster of variables related to the expected performance level; *criteria* as concepts related to the variable to be appraised.

Criteria operationalize the standard in terms of measurable quantitative guidelines. Criteria can be *general* in nature related to people (group) or situations (problems) or more *specific*, related to an individual person with a particular problem. The criteria are detailed and tailored to meet the needs of this person, see Scheme II — 1 (Fig. 1.3).

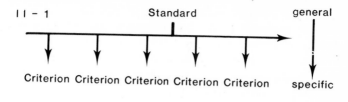

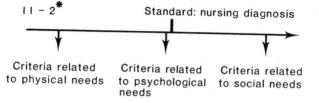

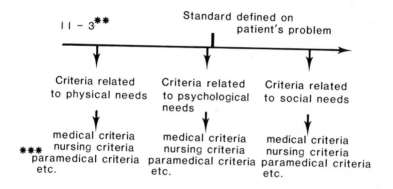

*evaluation of nursing care: mono- or intra-disciplinary
**evaluation of patient's problem related to care
    delivered, multi- or inter-disciplinary
***allied health professions

Fig. 1.3   Scheme II Standards and criteria

Example 1 of a standard

Each newly diagnosed responsive diabetic patient admitted to the medical ward of a general hospital is to be enrolled in a diabetic teaching program.

Criteria   *Assessment of need to learn*
              patient's knowledge of diabetes;

Assessment of readiness to learn
acceptance of handicap; the role of the family support system;
Setting of objectives
development of careplan, based on the needs of this diabetic patient;

Subcriteria   *Teaching — Learning*
introduction to the differences between insulins: brand names; teaching of the mechanics of using an insulin syringe; etc
Evaluation and reteaching if necessary

Example 2 of a standard

The care of each healthy mother and newborn baby delivered at home, is to be supervised by a qualified nurse/midwife for at least a period of ten days postpartum twice daily.

Criteria   Related to mother
Assessment of need to learn
knowledge of post-natal body changes in mother and baby;
Assessment of readiness to learn
perception of self in mother role; acceptance of baby; physical condition of mother; availability of support system;
Setting of objectives
teaching postpartum hygiene and self-care;
Teaching — Learning
focus on 1 ...
         2 ...
         3 ...
teaching care of baby;
focus on 1 ...
         2 ... etc.

Subcriteria   Related to baby
control of umbilical stump: hygiene, dressing; fluid balance; breastfeeding 6 times a day both sides; body position mother and baby; breast hygiene and massage; etc. micturition and bowel movements (meconium) amount, consistency, color, odor; etc.
Evaluation and reteaching if necessary.

In the above examples the standards have been formulated in objective, measurable terminology. This has not been done for the criteria so far. However each criterion should be written in the same way as the standard so that no misunderstanding about its function can occur. The description of criteria should be so clear that a nurse, other than the person involved in writing the criteria, is able to use them without the need to receive additional explanation about their meaning. The educational goals that guide the use of criteria are analogous to the phases of the nursing process.

Nursing process	*Educational goals*
Assessment: Identification of needs, Collection of information	Assessment: Need to learn, Readiness to learn
Planning: Setting of objectives, Development careplan	Setting of objectives
Implementation: Nurses' intervention by implementation of careplan	Teaching — Learning
Evaluation: Process and outcome	Evaluation
Follow-up: Are objectives reached? If not: reassessment	Are teaching goals reached? If not: reteaching (Redman, 1980, p. 25)

DEVELOPMENT OF STANDARDS FOR NURSING CARE AT NATIONAL LEVEL

It is possible to develop nursing standards at national or even at multinational level since professional nursing is based on the same underlying principles as defined by Henderson (1961):

> *Nursing* is to assist the *individual*, sick or well, in the performance of those activities contributing to *health* or its recovery (or to a peaceful death) that he would perform unaided if he had the necessary strength, will or knowledge. And to do this in such a way as to help him gain independence as rapidly as possible,

and stated in a comprehensive manner by nurses engaged in the development of nursing within the organizational context of the World Health Organization, Regional Office for Europe.

> *Nursing* is a fundamental activity carried out by all *individuals* and groups in society. In its organized form it is a discrete *health* discipline. It is both an art and a science. Its science is an organized body of knowledge gained through research and

analysis and its art is the creative use of that knowledge for the well-being of people. Nursing's primary responsibility is to assist individuals, families and *communities* to optimize physical, social and psychological function during varying states of health and at all stages of life. This means that the discipline is involved in functions which relate to health as well as to illness, and to any period of an individual's life from conception to death (WHO/EURO Nursing care of the elderly, 1981b, p. 2).

Both definitions cover the components of the nursing paradigm, i.e., people, health, environment and nursing. These components are interrelated and interdependent of each other. The interrelationship among the components determines the profile and dimensions of nursing as well as the directions of its development. It clarifies the boundaries of nursing (Thibodeau, 1983, p. 8).

Efforts have been made to develop criteria for routine nursing care. Aside from the question whether you can speak of routine in working with people, nurses have found themselves either in the process of writing standards or limiting the criteria to the description and standardization of nursing techniques. The evaluation of comprehensive nursing in terms of observable, measureable criteria is done most effectively at the patient care level, in the hospital or in the community.

The development of nursing standards has been more successful. It is the responsibility of the nursing profession to regulate, control and improve its practice. The self-regulation of nursing to assure quality in performance to the public is an authentic hallmark of a mature profession (ANA, 1980). However standards of care are not only a safeguard to the public, they also protect the nurse against malpractice and unprofessional conduct of nursing (Rychtelska, 1982).

It should be noted that the adoption of standards for nursing practice is not inherent to the introduction and use of these professional guidelines. Standards for nursing practice are used in those countries where professional performance is monitored for quality in order to be licensed and maintain accreditation (e.g., USA). Nurses in countries without or in process of developing an accreditation system use standards for nursing practice as guidelines in policy, planning and education.

Example 1
The American Nurses' Association published standards for nursing practice as early as 1973. These provide a basis for the

development of criteria and have been followed by a model for implementation (1976). A critical review of the standards showed strength in the operationalization of nursing as a discipline and the adaptation of the medical model. The standards could be improved by a redefinition of some constructs (Stevens, 1974). Nevertheless, the ANA standards have served as a model for the development of standards of nursing care in a number of countries and have been used as reference material in multinational conferences on quality of nursing care.

Example 2

The Canadian Nurses Association (1980) Task Group on National Standards for Nursing Practice developed a definition of nursing practice and related standards. These are clustered as follows:

1. A conceptual model for nursing
2. Nursing process
3. Helping relationships and
4. Professional responsibilities.

For each part are listed:

a criteria variable

a nursing standard

nursing behavior.

Example 3

In 1976 a national seminar on evaluation was organized by the *New Zealand Education and Research Foundation* (NERF) in Wellington, New Zealand. The concept of evaluation was discussed within the framework of structure, process and outcome. Kinross saw evaluation as a tool in the process toward independent nursing practice; Salmon pleaded for an evaluation system understood by nurses and other health workers and emphasized the importance of asking the right questions. Pitts regarded evaluation as a management tool. In the same year the NERF published nursing standards (Shetland, 1976).

Example 4

The *Royal College of Nursing* (RCN) published Towards Standards: A Discussion on Standards of Nursing Care. It presents the prerequisites for the professional control of standards of nursing care (1981).

Although the examples given are all related to English-speaking countries, we know that in many other places in the world nurses are concerned about the quality of professional practice and have made efforts to develop standards of nursing care. It became evident in the collaboration among nurses who are presently participating in the WHO/EURO Medium Term Programme in Nursing/Midwifery in the European Region, also during the conferences of the Workgroup of European Nurse Researchers. Why don't we have access to these standards or at least to the description of nursing care practice in those countries? The most serious barrier is *language*. Nothing is more complex to translate than the terminology of concepts, particularly in a field that is still so unexplored. It is such an important issue that a separate paragraph will be devoted to the meaning of language. Another reason is that nurses have been so conditioned to deliver perfect work that there is some reluctance to share an unfinished product that could still be improved. Nurses are not always aware of the help they could give each other in sharing ideas, experiences and writings, even in process of development.

Finally, standards of care emerge from knowledge of practice. The nurses involved in the groundwork of its development are not always the persons whose contribution is accredited at levels where the sharing and exchange of new knowledge takes place. Therefore it is so important that nursing as a practice-oriented profession safeguards the liaison between the clinical area in the community and in health care institutions, educational settings and the policy level (Henderson, 1980).

Examples of such liaisons are:

— The nursing practice research unit of Northwick Park Hospital and clinical research center, Harrow, England;
— Rush-Presbyterian/St Luke's Medical Center, Chicago, Illinois, USA;
— The integrated Nursing Research Project of the Research Unit of Hospital Administration and Medical Care Organization of the Catholic University in Louvain, Belgium.

THE MEANING OF LANGUAGE IN THE EVALUATION OF NURSING CARE

Evaluation of nursing care requires comprehensive knowledge and understanding of a common language and agreement on the meaning of quality concepts and terminology. Since the nature of nursing practice frequently forces nurses into thorough, but quick decision making, the importance of the exploration and verification of the appropriate terminology is often underestimated. New concepts are sometimes adopted without careful scrutiny of their meaning and application to nursing practice. Examples are concepts like nursing process, primary nursing, and nursing audit. Ask a group of nurses to define each of these concepts and the result is a wide variety of opinions.

Language is an important means in the communication among people. The proper use of language facilitates the exchange in information and the development of common goals and objectives with regards to the evaluation of nursing care. This applies in the first place to situations where nurses are working together in any health care setting in one country. Most evaluation will take place within the boundaries of one particular hospital, a community health center or professional organization (Hockey, 1979).

However the expansion of multinational exchange and collaboration has intensified the contacts among nurses from many nations. Professional meetings serve as a platform for discussion and dialogue where people share, learn together and make decisions on health and nursing issues of current interest that have impact on the planning of services and education of the future. Examples are the regional meetings of the member Associations of the International Council of Nurses, the World Health Organization and a wide variety of nursing speciality groups.

The planning and evaluation of programs at multinational level demand a sound knowledge of the content matter and a mastery of the language of communication. For nurses whose mother tongue is other than English, Spanish, Russian, to name a few examples, language can be a barrier in expressing the true meaning of words and symbols. Some concepts cannot be translated because they are unknown in certain cultures or are given a different meaning than is familiar to the nursing community at large. Examples are the use of the word nursing, for which there is no direct translation into Swedish, and the description of the concept assessment in other languages that has even intrigued linguists. In limiting ourselves to the use of one or two languages, we exclude unwillingly a large population of nurses from other origins, and exclude ourselves from resources emerging from foreign cultures and published in other parts of the world. The mastery of more than one language is an important step to promote multinational exchange. However knowledge of languages should be complemented by understanding of the significant characteristics of the culture and its people (Edgerton et al, 1974, pp. 38–39; Harris & Moran, 1979, pp. 78–81).

The use of language in the evaluation of nursing care has gained in significance since many of the theories in nursing and the methods of investigation are rooted in the Anglo-Saxon cultures.

EVALUATION IN ITS APPLICATION TO NURSING PRACTICE

Evaluation is a continuing process throughout the delivery of health/ nursing care. It has been defined as follows:

> Evaluation of health services: the systematic process of determining the extent to which an action or sets of actions were successful in the achievement of predetermined objectives (WHO/EURO, 1979, p. 3).

This definition applied to nursing is:

> Evaluation of nursing care is the process of scientific reasoning in order to determine the extent to which the results or outcomes of nursing interventions were successful in achieving predetermined standards of nursing care.

Evaluation is a process in which each professional nurse is engaged regardless of the setting of practice and the resources available. The main instrument of evaluation is the nurse's attitude towards practice and accountability developed throughout professional education preparing her for critical analysis and problem solving.

The use of techniques and methods is the structure within which nurses can discuss the process and outcomes of nursing care with their peers. Familiar to many nurses is the method of scientific inquiry or the nursing process, a concept more clearly described as the process of logical reasoning. It is an effort to provide individualized, personalized care to the patient/client (Henderson, 1982). The principles are based on problem identification, analysis and problem solving. The *process* applies in the same way to any professional practice. However the term *nursing* process is used when the problem under scrutiny is evaluated from a nursing perspective.

Evaluation of professional care delivered to patients/clients (Scheme III–1)

Evaluation of nursing care can be developed as a monodisciplinary activity. This means that nursing as a discipline sets nursing standards and criteria. The characteristics are that the patient/client needs for health care are examined from a nursing perspective and that these needs are met by employing nursing resources, Scheme II–1; II–2 (Fig. 1.3).

Although the other health professions, analogous to nursing, also set standards and criteria to evaluate their clinical performance, it does not mean that the care delivered to the individual is fragmented into nursing and other parts. On the contrary, it illustrates an awareness of professional responsibility and a more conscious use of expert knowledge from within the discipline.

However evaluation of nursing care does not have a function in situations in which nurses question nursing and feel insecure about the validity of the body of nursing knowledge reflected in the way nursing is practiced. If a nurse questions her identity as professional and care provider it will be very difficult, if not impossible, to define the quality of nursing care that ought to be accomplished.

Evaluation of patient/client needs related to professional expertise (Scheme III–2)

Another approach is to examine the care from a perspective of patient's needs and problems. This is a joint responsibility of all health professionals and can be planned in a multidisciplinary approach.

The focus of attention is the help-seeking person to whom a variety of resources are available, delivered by different health professionals who complement each other's expertise. The professional goals of each of the disciplines are subordinate to the common goal, the service to the patient/client. Each discipline is still responsible for the development of professional standards and criteria, but prepared to integrate these together with the quality measures set by other disciplines into one masterplan, Scheme II–3 (Fig. 1.3).

This model is effective when people are willing to work as a team and accept each other's contribution on the basis of *equality* and mutual *respect*. The multidisciplinary approach to evaluation of care requires give and take on the part of all of its participants.

Examples of evaluation

    Example 1
    Evaluation based on professional expertise without the availability of technical resources for quality monitoring:
    A community health nurse who is working as an all-round

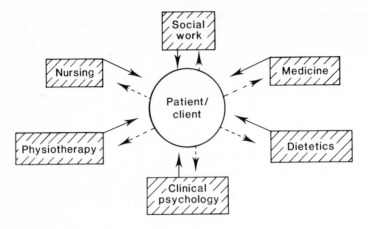

Scheme III 1. Evaluation of professional care delivered to patients/clients

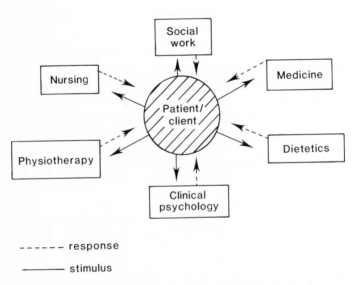

- - - - - response

———— stimulus

Scheme III 2. Evaluation of patient/client needs related to professional expertise.

Fig. 1.4

health worker in a rural area of her country, questions the causes of the high infant mortality rate. Evaluation on basis of knowledge, insight and experience in working with the local population makes her assume that the mortality is related to life-styles. She decides to promote the improvement of sanitation and nutrition and focuses on community teaching. The infant mortality decreases over the years. A justified assumption is the relationship between the delivery of health-oriented nursing services and the lowering of infant mortality rate.

Example 2

Evaluation based on professional expertise and the access to a quality monitoring system.

A nursing home nurse has observed that some frail elderly, incontinent residents are able to control bladder and bowel function from the moment they become involved in meaningful recreation and intellectual stimulation. In an effort to explore cause and effect it is decided to review a number of patients' records, to examine care-plans and to observe residents' behavior. This is complemented by interviews. The evaluation of the observed phenomena is used as baseline information for the development of nursing intervention.

This approach may seem logical and based on common sense; there are some pitfalls however. The most serious problem is the inability of nurses to practice nursing (Henderson, 1980; Hall & van Maanen, 1981; McFarlane, 1981). One of the characteristics of professional nursing is that the nurse is fully responsible and accountable for the clinical decisions made with regard to the health and well-being of the *individual* patient/client. If the nurse is assigned to a group of patients/clients this unique element gets lost in the fragmentation of nursing care.

Another problem is that most nurses are employed by health care institutions or community organizations. This means that in general the responsibility for nursing services is laid in the hands of the nursing director. Although the *nurse-manager* cannot be made responsible for the many decisions taken by the nurse-clinician with regard to the nursing care for individual patients, she is *accountable* for *all* the *nursing* services rendered. This is especially the case in countries where nursing education has not entered the system of higher education and where nursing practice is not covered by updated legislation.

Since the quality of care and the level of evaluation are related to the competence and ability of the nurse to practice professional nursing, let us compare the characteristics of a profession to the models of task-oriented, team and primary nursing.

Characteristics of a profession applied to nurses and nursing practice

Professional authority

visualized in professional behavior is attitude and action that characterize nursing. Its foundations are education, autonomy and service. It is the connecting thread between attitude and practice which are characterized by a philosophy of nursing. It is demonstrated through accountability for behavior to client and inevitably to society as well as by responsibility for practice with continually updated knowledge and skill (College of Mount St. Vincent, 1979).

A systematic body of knowledge

this is the knowledge gained by intellectual preparation, the construction of theory via research and in most instances the prolongation of education.

Community sanction

the profession, acknowledged by society, takes control of its education, accreditation and licensure which are regulated by legislation.

Ethical codes

they are the ethical guidelines for practice and guarantee a certain level of quality performance to the public and protect both the public and the professional against malpractice.

Professional culture

it is the social climate in which the professional functions and seeks his contacts. It is characterized by certain values, norms and symbols (Greenwood, 1957).

These characteristics of a profession applied to present nursing can be shown in profile form.

Professional authority

The *task*-oriented nurse is responsible for the delivery of quality nursing care, but is not given the authority to initiate major changes and adjustments. The lack of formal authority does not mean that the nurse has no informal power, gained by clinical experience and her role in relation to staff and recipients of care.

In *team* nursing, the individual nurse may have gained some authority because of expertise in team leadership, but the impact is limited to the direct relationship recipient-nurse and from the nurse to other team members.

The *primary* nurse, who is responsible for the delivery of individualized care, is given more authority. It will depend on the clinical setting and the country how the boundaries of independent nursing practice are defined. Primary nursing has been acknowledged by the nursing profession, but it is of too recent origin to have gained the recognition of society other than on an individual basis.

Body of knowledge

Task-oriented nursing is rooted in apprenticeship, the learning on the job. The existing knowledge is dormant. It has not been identified and thus not been integrated in nursing practice. Nurses recognize that there is more to nursing than the performance of tasks, but have not gained the knowledge and skills to clarify concepts and processes.

In *team* nursing, there is a growing awareness of the significance of nursing care for the health and well-being of people. The approach to the development of knowledge is deductive (theory → practice). Nurses use the knowledge and contribute experiences to the construction of new knowledge, but it is developed by others.

Primary nursing is based on the active use of the body of knowledge in nursing and the recognition of nursing as an emerging human science. It is expected that the primary nurse is able to assist in the development and testing of new knowledge by exercising the highest attainable level of practice. The development of new knowledge is grounded in a deductive and inductive approach (practice → theory).

Community sanction

Education. The quality of nursing is determined by the level of basic education and the conditions to practice nursing. First level nurses are prepared through programs that vary worldwide from training in hospital diploma schools to education at academic levels offered in universities or colleges of higher technical education. In countries with predominant hospital based nursing preparation, nurses have limited control over their education in contrast to other countries where academic nursing, taught in institutions for higher education, has resulted in a higher degree of self-governance. Although the level of education is no absolute determinant for the level of performance, it cannot be denied that the capability of the graduate nurse is related to the quality of education and professional preparation.

The well-educated nurse feels less attracted by task-allocation.

> A degree and university life tends to make one unsatisfied with the routine and discipline of nursing and especially the medical hierarchy — though not lessening the satisfaction of the actual work with patients (House, 1975).

The purpose of education is to prepare nurses who are qualified to give high quality *health* services to the public from a nursing perspective. A legitimate question is whether high quality education has resulted in improvement of patient/client services.

A survey in 25 countries of the European Region showed that in 10 countries the Ministery of Health was solely responsible for basic nursing education (nursing manpower), whereas in 8 countries the authority was shared between the Ministry of Health and the Ministry of Education (professional preparation). The UK is the only European country where nurses have control of nursing education regulated via the General Nursing Council for England and Wales (WHO/EURO, 1981a).

Accreditation. In some countries graduation from a basic nursing program qualifies the nurse automatically to practice nursing, e.g., the Netherlands. The qualification is based on final examinations, offered by the School of Nursing, or in a standardized form, by the responsible Ministry or comparable authority; the situation differs by country. In other countries nurses have to take a licensure examination before they are legally authorized to practice nursing, e.g., USA. Interestingly enough the licensure examination does not distinguish between the different levels of professional competency upon graduation. The development of standards for nursing practice is a professional responsibility. Standards for nursing care reflect a philosophy of nursing; nurses' beliefs and values are described in terms of nursing interventions. The definition of standards may be activity or task directed at the beginning, gradually evolving to a more sophisticated professional level.

Legislation. An effective functioning of nursing services has its basis in adequate legislation. In the majority of countries nurses function legally under the direction or supervision of a physician. The legal definition of the nurse adopted by the International Council of Nurses (1975) does not accord with the reality, which places the nurse in a subordinate role in relation to the physician (WHO/ EURO, 1981). Although most countries have some form of nursing legislation, the laws often do not keep pace with the advancements of the profession, in fact can be restrictive with regard to expansion of nursing. Illustrative are the following trends:

1. The education and practice conditions for nurses are focused on *curative* medicine rather than on any well developed understanding of nursing.
2. There is an increased provision in legislation for the transfer of nurses from one basic nursing group to another, e.g., from psychiatric to general nursing.
3. The difference between first and second level nursing personnel is ill-defined (WHO/EURO, 1981).

In the USA nursing practice is regulated by Nurse Practice Acts, that allow the qualified nurse to work as an independent practitioner (Bullough, 1983; Conway, 1983; Monnig, 1983).

Ethical code

The Code of Ethics, adopted by the International Council of Nurses (ICN) in 1975, applies to the nursing practice of all level nurses. Nurses are frequently confronted with ethical dilemmas and have taken the lead among the health professions to bring these problems to the attention of the nursing community and the public. Although nurses have given evidence of their expertise and concern for people by the identification of ethical issues, the answers to complex problems require a profound knowledge and interaction with professional and lay people. Ethical dilemmas in nursing care are intertwined with the care provided by other health professions and should be studied in a collaborative effort. However each profession has its own responsibility and is accountable for its decisions. Nurses have taken a stand in cases where quality of life of patients was at risk. The impact of such action is related to the level of professionalism and the legislation covering nursing practice (ICN, 1977).

Table 1.3 The characteristics of a profession related to the performance of nursing practice and the potential for self-governance of nursing by nurses

Task oriented nursing	Team nursing	Primary nursing
Body of knowledge		
existing, but not identified: Is nursing a profession? What is its identity?	increasing awareness of a body of knowledge; consumer of this knowledge	creative awareness of the body of knowledge and emerging nursing science; participant in the development of new knowledge (research)
Professional authority		
is absent although the recipient of nursing care is subject to nurse's judgement; the patient/client is considered unable to evaluate the quality of nursing care	professional nursing competence is recognized on an individual basis; the recipient of care is invited to evaluate the quality of nursing, but the impact of this feed-back is limited to micro-level nursing	professional nursing competence is acknowledged by the profession and the public; the evaluation of nursing care is a professional responsibility; the contribution of the recipients of care (public) initiates quality improvement
Community sanction		
education		
nursing education is related to training for the job; the control falls under the jurisdiction of non-nurses with few exceptions	nursing education is in transition; the preparation for nursing practice embodies aspects of training and professional education; nurses are gaining responsibility for nursing education	professional education prepares the nurse for professional practice; responsibility for education is under the jurisdiction of the profession of nursing
accreditation, licensure: varies by country		
standards for nursing practice are function (job) related	standards for nursing practice are function and profession related but still with emphasis on the job	standards for professional performance are developed and controlled from within the profession
legislation: varies by country		
nurses function legally in a subordinate role		professional practice is regulated by law; Nurse Practice Acts
Ethical code		
existing, but not functional because of lack of automony	closer relationship to patient/client confronts the nurse with ethical problems and dilemmas	professional practice is rooted in an ethical code that guides performance

Professional culture		
socialization in non-professional occupations with adoption of norms, values and symbols	a vocation in transition to profession; changing norms, values and symbols	socialization in professional environment among the nursing profession and to other (health) professions

The attributes of a profession: Greenwood (1957); Styles (1982); WHO EURO (1979); compiled by van Maanen (1983)

Professional culture

> Socialization is the process of inculcating new values and behavior appropriate to adult positions and group membership (Rosow, 1974, p. 31).

Nursing, viewed from an international perspective, is in a phase of transition from a vocation to a profession. Nurses socialize in non-professional as well as in professional groups, depending on their educational background, position and social status. Nurses are encouraged to take on new roles in the process of professionalism.

Although role transition is an internal process of development in the first place, new roles are shaped within the broader context of the society. Styles (1982, p. 8) discusses professionalism as a process that will only be achieved through the professionhood of its members.

Professionhood is the concept that applies to the characteristics of the individual nurse as member of the nursing profession. Knowledge of self facilitates the diaglogue with others in the process of professionalism. Professionhood is a foundation of professionalism, however a profession gains its recognition by the unity of its members in one professional group (Fields, 1981).

CONCLUSIONS AND RECOMMENDATIONS

If we accept that the evaluation of quality of nursing care is a professional responsibility, it is implicit that the conditions to practice professional nursing should be present. According to the definitions of evaluations of nursing care and primary nursing, practice should be based on a philosophy reflected in the quality care provided to patients/clients and their families. The instrument is the process of logical reasoning.

The most significant model of professional practice is at present primary nursing, characterized by:

— a *philosophy of nursing*, derived from the existing body of nursing knowledge and research
— the use of *process of logical reasoning*
— *professional authority*, awarded on the basis of education, autonomy and nursing excellence in services rendered to the public; the nursing care provided is continuous, co-ordinated, comprehensive, individualized and patient/client centered; human rights are respected; the recipient of care is challenged to become a participant in the planning and decision making with regard to his health and well-being
— *collaboration* with other professions in the improvement of quality health services, the use of new knowledge and science and in balancing the advances of technology versus the quality of life of people.

Professional practice is regulated by legislation covering education, licensure and practice.

If primary nursing is accepted as a model of professional practice, which embodies the responsibility and accountability of the individual nurse for the evaluation of nursing care provided, what about other settings where nurses practice under less favorable conditions? Are the consequences that quality nursing care can only be delivered in health care institutions and community nursing organizations that are staffed by professional nurses? Does it mean that when there are no professional nurses that no quality nursing care is practiced?

The answer is up to the reader, who can judge best whether the observed nursing practice is meeting her/his standards of professional nursing excellence. An analysis of the present situation is the starting point of change and development. Is not each professional nurse a potential change agent?

A few comments should be made.

— Although education is an important condition for quality improvement, it does not guarantee that the well-educated nurse is equally competent in the clinical area, where evaluation of nursing care takes place. Although one cannot practice what one does not know, most learning takes place *outside* formal educational settings in the working environment, at home and in the community. It is well possible that a nurse through training and experience accomplishes a level of professional expertise that is the equivalent of primary nursing, professional practice.
— Quality nursing care is based on a philosophy of nursing and not inherent to the existence of a quality monitoring system, in spite

of the fact that it may facilitate the conditions under which nursing excellence is accomplished.

— The role of the individual nurse is of crucial importance for the maintenance and improvement of quality nursing care. However, the professionhood of the individual should be profiled within the nursing community, united in one profession. The conditions for quality assurance, education, legislation and practice are developed most effectively from within the profession as a bargaining power, in collaboration with health authorities that legitimize those changes. The profession is responsible for the development of the conditions to practice nursing.

An example of collective action in nursing is the WHO/EURO Medium Term Programme in Nursing/Midwifery in Europe. It is a development and evaluation project based on planned change and action research. Its major objectives are to strengthen the nursing component of all health services and to assist countries in the development of conditions under which nurses can practice *nursing* in institutions as well as in community services. Nurses from a selected number of countries have participated in the construction of assessment and evaluation tools that after thorough testing will be implemented in nursing practice throughout the region. A priority is at present to familiarize nurses with the method of logical reasoning by implementing the nursing process. An important outcome of the

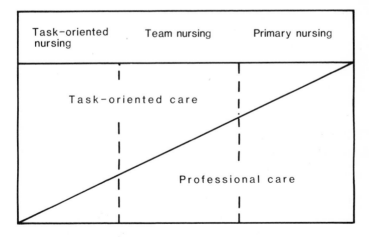

Fig. 1.5   Scheme IV

program so far has been that nurses, in a joint effort, have mobilized their expertise and developed strategies to help their colleagues to accomplish professional nursing goals.

The development of evaluation programs can be structured at various levels. It can be approached from: a micro-level, patient/client oriented; a meso-level, directed toward nursing services and from a macro-level when nurses together with other disciplines evaluate nursing within the context of health services (Bergman, 1980).

Table 1.4 Examples of evaluation tools at the clinical level in sequence development

Task oriented nursing	Team nursing	Primary nursing
Records (medical record) nursing reports related to the *group* of patients	(medical record) nursing reports per *individual* patient/client	(*patient* record) used by all the health professionals involved in patient care
medical orders are carried out	medical and *nursing* orders become part of patient care-plan	the care planned by nurses, physicians and other disciplines is *integrated* in one care-plan
no review other than for medical purposes	nursing review; *process* evaluation	nursing review; *process* and *outcome* evaluation; peer-review
Planning *work* conferences allocation	team conferences combination of *task* assignments and *patient* conferences	*patient/client* conference attended by nurses only or nurses and other team members or nurses and other team members including the patient/client
Inservice education non-existent or *medical*, technical oriented	medical, technical oriented; nursing issues may arise as part of *psychosocial* care	*nursing* oriented; medical focused as far as relevant for care planning; the patient/client may be included as participant and 'instructor'
Quality monitoring nursing non-existent in structured manner	informal monitoring may develop, initiated by *nurses* and structured by *others*	*nurses* take *control* of *nursing* practice and develop systems of quality monitoring
Quality monitoring hospitals (and other health care settings) nurses follow orders	nurses work with others, nursing component not clearly identified	nurses are responsible for the development of nursing *standards* and *criteria*

Each program encompasses the use of certain evaluation tools that can vary in sophistication throughout the transition from vocational to professional nursing.

These examples (Table 1.4) only serve as indicators in a process of development. Phases in the transition period may not occur, may overlap or take on a different route of development. The function of the scheme is to assist nurses in their thinking about change in a process toward professional practice.

In the development of evaluation of nursing care programs, nurses 'need to look for both universal and specific cultural nursing standards in education and practice to advance the discipline of nursing' (Leininger, 1980, p. 124).

Evaluation of professional practice should in the end serve the public, because their health care is our concern.

REFERENCES

American Nurses' Association 1976 Standards: Nursing practice. ANA, Kansas City

American Nurses' Association 1980 Nursing: a social policy statement. ANA, Kansas City

Annel M V 1982 A Guatemalan experience: The community, not the doctor in control. World Health Forum 3: 282–284

Bandman E L, Bandman B 1981 Health and disease: A nursing perspective. In: Caplan A L, Engelhardt H Tristram Jr, McCartney J J (eds) Concepts of health and disease, interdisciplinary perspectives. Addison-Wesley, London, p 667–693

Baumgart A J 1982 Nursing for a new century — a future framework? Journal of Advanced Nursing 7(1): 19–23

Berg H V 1974 Nursing audit and outcome criteria. Nursing Clinics of North America 9(2): 331–335

Bergman R 1980 Evaluation of nursing care — could it make a difference? In: Lorensen M (ed) Collaborative research and its implementation in nursing. The Danish Nurses Organization, Copenhagen, p 103–115

Boekholt M 1982 Verandering van zorgconcept kan diep ingri jpen in de structuur en inrichting van de zorgverlening. (Translation: Change in the concept of care can have great impact on the structure and planning of patient care). Het Ziekenhuis 12(6): 276–280

Bullough B 1983 The relationship of Nurse Practice Acts to the professionalization of nursing. In: Chaska N L (ed) The nursing profession, a time to speak. McGraw-Hill, London, ch 45, p 609–633

Canadian Nurses Association 1980 Development of a definition of nursing practice. The Canadian Nurse 76(5): 11–15

College of Mount Saint Vincent 1979 Self-study report of the Department of Nursing. College of Mount Saint Vincent, Riverdale, New York

Conway M E 1983 Prescription for professionalization. In: Chaska N L (ed) The nursing profession: a time to speak. McGraw-Hill, London, ch 2, p 29–38

Crow R 1981 Research and the standards of nursing care: What is the relationship? Journal of Advanced Nursing 6(6): 491–496

Donabedian A 1976 Measures of quality nursing care? Experts agree valid approach not yet found. American Journal of Nursing 76(2): 186

Edgerton R B, Langness L L 1974 Methods and styles in the study of culture, Chandler and Sharp, San Francisco

Fields H E 1981 Study of professional behavior with an emphasis on professional nursing education and practice. Unpublished dissertation, Columbia University Teachers College, New York City

Greenwood E 1957 Attributes of a profession. Social Work 2(4): 44–55

Hagen E 1976 Conceptual issues in appraising quality of nursing care. Unpublished paper, Columbia University Teachers College, New York City

Hall D 1979 A position paper on nursing. World Health Organization, Regional Office for Europe, EURO/NURS/75.1 Rev. 1, Copenhagen

Hall D, van Maanen H M Th 1981 Health manpower, the example of nursing. In: Health services in Europe, 3rd edn. Volume 1: Regional Analysis. World Health Organization, Regional Office for Europe, Copenhagen, p 71–77

Harris P R, Moran R T 1979 Managing cultural differences, Gulf Publishing, London

Hegyvary S T 1982 The change to primary nursing, a cross-cultural view of professional nursing practice. Mosby, St Louis

Henderson V 1961 Basic principles of nursing care, International Council of Nurses, London, p 42

Henderson V 1978 The concept of nursing. Journal of Advanced Nursing 3(2): 113–130

Henderson V 1980 Preserving the essence of nursing in a technological age. Journal of Advanced Nursing 5(3): 245–260

Henderson V 1982 The nursing process — is the title right? Journal of Advanced Nursing 7(2): 103–109

Hockey L 1979 Collaborative research and its implementation in nursing. In: van Maanen H M Th (ed) Collaborative research and its implementation in nursing. Nationaal Ziekenhuis Instituut, Utrecht, 83–94

Hoenson F 1972 Het zebraped (Translation: The cross-over) position paper written for the Municipality of Amsterdam

Hogarth J 1978 Glossary of health care terminology, Public Health in Europe 4, World Health Organization, Regional Office for Europe, Copenhagen

House V G 1975 Paradoxes and the undergraduate student nurse. International Journal of Nursing Studies 12: 81–86

International Council of Nurses 1975 Code of ethics. ICN, Geneva

International Council of Nurses 1977 The nurse's dilemma, ethical considerations in nursing practice. ICN, Geneva

Koene G, Grypdonck M, Rodenbach M Th, Windey T 1980 Integrerende verpleegkunde (Translation: Integrated nursing). Wetenschap en Prakti jk. Lochem, De Tijdstroom

Lacronique J F 1981 Evaluation of health care. In: Health services in Europe, 3rd edn. Volume 1: Regional analysis. World Health Organization, Regional Office for Europe, Copenhagen, 51–57

Lanara V A 1981 Heroisim as a nursing value, a philosophical perspective, Sisterhood Evniki, Evangelismos Hospital Athens

Lanara V A 1982 Development of a scientific foundation of the nursing profession. In: Lerheim K (ed) Collaborative research and its implementation in nursing, The Norwegian Nurses Association, Vettre, 98–103

Leininger M 1980 Reflections on the development and implementation of standards for nursing practice. In: Lorensen M (ed) Collaborative research and its implementation in nursing. The Danish Nurses Association, Copenhagen, 123–138

Lorensen M 1979 Development of quality criteria for nursing practice. In: van Maanen H M Th (ed) Collaborative research and its implementation in nursing. Nationaal Ziekenhuis Instituut, Utrecht, 111–118

MacPhail J 1972 An experiment in nursing: planning, implementing, and assessing planned change. Case Western Reserve University, Frances Payne Bolton School of Nursing, Cleveland

Marram G, Barrett M W, Bevis E O 1979 Primary nursing, a model for individualized care, 2nd edn. Mosby, St Louis

Mayers M G, Norby R B, Watson A B 1977 Quality assurance for patient care, nursing perspectives. Appleton-Century-Crofts, New York City

McAlary B 1981 The reliability and validity of hospital accreditation in Australia. JANFORUM Journal of Advanced Nursing 6(5): 409–411

McFarlane of Llandaff J 1981 Standards of nursing care, how can research help the nurse manager? In: Lanara V (ed) Collaborative research and its implementation in nursing. The Hellenic National Graduate Nurses' Association, Athens

Monnig R L 1983 Professional territoriality in nursing. In: Chaska N L (ed) The nursing profession: a time to speak. McGraw-Hill, London, p 38–49

New Zealand Nursing Education and Research Foundation (NERF) 1978 Papers presented at a national seminar on evaluation, Massey University. Speakers: Kinross N J, Salmon B, Pitts A. NERF Studies in nursing no. 6, NERF Westbrook House, Wellington

Padilla G V, Grant M M 1982 Quality assurance programme for nursing. Journal of Advanced Nursing 7(2): 135–142

Peccei A 1983 Interview with Kees Caljé, (ed) NRC Handelsblad, overseas edition, Peccei (President of the CLUB OF ROME) Only culture can save the world: If we don't learn how to cope with technology, we will destroy ourselves

Redman B K 1980 The process of patient teaching in nursing, 4th edn. Mosby, London

Rogers M E 1970 An introduction to the theoretical basis of nursing, Nursing Science 1. Davis, Philadelphia

Rosow I 1974 Socialization to old age. University of California Press, Los Angeles

Roy C 1976 Introduction to nursing: an adaptation model. Prentice-Hall, Englewood Cliffs

Royal College of Nursing 1981 Towards standards: a discussion document. Second report of the RCN Working Committee on Standards of Nursing Care, RCN, London

Rychtelska M 1982 The ILO's international standards: What they mean for nurses. International Review 29(5): 136–138

Scott Wright M 1983 Present realities, future strategies, Keynote address Friends of the School Annual Lecture, University of California, San Francisco

Shetland M L 1976 Nursing standards, who, why and how? New Zealand Nursing Education and Research Foundation, Wellington

Smit A 1981 Health in Europe. In: Health service in Europe, 3rd edn. Volume 1: Regional analysis. World Health Organization, Regional Office for Europe, Copenhagen, p 3–11

Stevens B 1974 ANA's standards of nursing practice: What they tell us about the state of the art. Journal of Nursing Administration 4(5): 16–18

Styles M M 1982 On nursing, toward a new endowment. Mosby, St Louis

Thibodeau J A 1983 Nursing models: Analysis and evaluation, Wadsworth Health Sciences Division, Monterey, California

van der Velden B 1983 Herbebossing mislukt, erosie gaat door, Ethiopie hongert. (Translation: reforestation failed, erosion continues, Ethiopia starving) NRC — Handelsblad, overseas edition

van Eindhoven J M B 1979 Een systeem van eerst verantwoordeli jke verpleegkundige. (Translation: a system of 'first responsible nurse') Lochem, de Tijdstroom

van Maanen H M Th 1979 Quality of nursing care: some general thoughts on the theoretical concepts. In: van Maanen H M Th (ed) Collaborative research and its implementation in nursing. Nationaal Ziekenhuis Instituut, Utrecht, p 107–109 van Maanen H M Th 1979 The nursing profession: ritualized, routinized or research-based? JANFORUM Journal of Advanced Nursing 4(1): 87–89

Wiener C, Fagerhaugh S, Strauss A, Suczek B 1980 Patient power: complex issues need complex answers. Social Policy 11(5): 30–38

Wiener C, Fagerhaugh S, Strauss A, Suczek B 1982 What price chronic illness? The politics of sickness. Society 19(1): 22–30

World Health Organization, Regional Office for Europe in Copenhagen 1978 Third Liaison Meeting with Nursing/Midwifery Associations on WHO's European Nursing/Midwifery Programme. Report

World Health Organization, Regional Office for Europe in Copenhagen 1979 Evaluation of inpatient nursing practice. EURO Reports and Studies, 4

World Health Organization, Regional Office for Europe in Copenhagen 1980 Fourth Liaison Meeting with Nursing/Midwifery Associations on WHO's European Nursing/Midwifery Programme. Report

World Health Organization, Regional Office for Europe in Copenhagen 1981a Legislation concerning nursing/midwifery services and education. EURO Reports and Studies, 45

World Health Organization, Regional Office for Europe in Copenhagen 1981b Nursing care of the elderly. WHO/AGE 81.1 6158B UNEDITED, Copenhagen

World Health Organization, Regional Office for Europe in Copenhagen 1982 Final Report on the workshop on the nursing process

World Health Organization Geneva 1974 Community Health Nursing. Report of a WHO Expert Committee Technical Report Series 558, Geneva

World Health Organization/UNICEF Geneva 1978 Primary health care. Report of the International Conference on Primary Health Care, Alma-Ata, USSR, Geneva

United Nations 1982 Report of the World Assembly on Aging. A/CONF.113/31, New York

Knowledge for practice: the state of clinical research

THE CURRENT STATE

For a number of years, nursing literature has reported on the lack of clinical research and, even more important, a lack of implementation of existing research findings. Nursing leaders verbalize a breach between research, education, clinical practice and the impact of the work environment. Nurses reiterate that if nursing doesn't do something, others will. These factors are combining to push nurses to incorporate research findings in daily practice for the purpose of measuring and improving the quality of patient/client care. To implement different practices based on research requires changes in attitudes of employers, employees, patients/clients and co-workers in both professional and support fields. Different practices require learning different skills and forgetting years of habit. Different practices necessitate acquiring a body of new knowledge. These changes will only become reality when and if someone really cares — for it is the caring person who dares to act to preserve that which is good, to revitalize that which has been spent and who derives satisfaction from innovation.

Kinds of research

For practical purposes research can be considered as historical, descriptive or experimental in nature. There is a place in nursing for all types.

Historical research tells us what happened in the past. From that information one can select positive factors, including patterns of approach or methods of practice that may still be effective.

Descriptive research reveals 'what is' at a given point in time. This form of research may be used to obtain opinions of people about the quality of service provided, a program planned, current beliefs about nursing and the needs of people or a myriad of other topics. Descriptive research can reveal the way an agency, any of its departments, or a group of agencies function. The description of the activities can provide information as to whether or not action is directed toward accomplishment of goals. In addition it can depict how much progress has been made toward achievement of aims. Descriptive research is a 'still' picture extracted from a continual moving film strip.

Experimental research is designed to gather information which will substantiate the current approach to the situation under study or will provide evidence that a different method will produce more effective results. Experimental research includes identifying the concern (problem), formulating hypotheses, developing a methodology, conducting the test, collecting raw data, collating it and using a test to determine the reliability of the information produced. It is inherent in experimental research that tools be tested for validity and reliability. A research experiment is one that is controlled. This means one group of subjects is exposed of the influence (effect) of an independent variable (treatment item) while another group is not. The exposed group is referred to as the experimental group; the other is the control group. Experimental research endorses the use of pretesting and post testing — determining certain data prior to and following exposure to the treatment.

All research is inextricably interwoven with questions of ethics, values, confidentiality and human rights. This paper will not explore these questions but the writer will assume that nurses participating in research — generated by nursing or other disciplines — will carefully consider those aspects prior to committing themselves to involvement in the projects.

THE NEED FOR CLINICAL RESEARCH

The providers

Nursing activities have to do with care rather than cure of people. These activities do not require a physician's prescription and many times a physician is relatively unaware of how nursing techniques are performed. Methods used by nurses in providing this care probably will remain static or even may regress and certainly will become invalid if nurses fail to research them. Some current methods are steeped in tradition and, having been passed from one generation of nurses to another, are inadequate. Sometimes the routines do not have any rhyme or reason, except that they have always been done 'this way'. Some nursing activity is learned by imitation and taught with very little if any basic underlying principles.

The care provided by a nurse is markedly influenced by the values and beliefs of that nurse. The initial forming of beliefs is generated by the family or the people performing this role. This source of beliefs (tradition, authority) will accompany one throughout life — the actual persons in the authority role changing. This core of beliefs is expanded as one matures. The situations — both planned and incidental (accidental) — which one experiences contribute to the development of one's philosophy.

The beliefs are directly affected by the value system of the individual. Value systems are engendered by the decade into which one is born and developed throughout the life span. The places one lives and the values of the parents, as influenced by ethnic ancestry, contribute to the formation of one's value system. The educational systems in which one finds or enrolls oneself contribute to the values developed as persons within a system and are permeated by its values. The values placed on religious practices and in the political ideologies, as well as the degree of involvement in each, contribute to the development of one's values. The economic status of the society, the family and the individual influences the values which develop. Values are also molded by the kind of job or vocation chosen by the individual and the practices of peers and associates. The acquisition of knowledge, especially that derived from scientific investigation (research) stimulates one to review personal beliefs and modify, reverse or confirm them.

Research is the most effective method designed for finding unity, or order of relationships, so that reliable guides for conduct can be developed. Many professions, businesses, and large industries make use of scientific inquiry in order to solve the identified problems, to find new material, to test the reactions of the public and the reactions of those people who are providing the service, conducting the business, etc. It would seem logical that nursing, which is attempting to become a profession and indeed in some states and countries is considered to be a profession, would in turn use scientific inquiry in the exploration of its actions and solution of its problems. Also nurses would make use of scientific inquiry as a basis for carrying out programs, not only of basic education, but of continuing education.

In Saskatchewan in the past, the majority of energies for raising the quality of care have gone indirectly into improving the basic preparation for nursing, learning how to recruit and to a lesser degree on the preparation of administrators and teachers. The emphasis on preparation for teaching has outweighed the emphasis on preparation for administration. Both have far exceeded emphasis upon the preparation of the clinical practitioner beyond the basic level. However, the time seems right and the opportunity seems to be available for nurses now to make use of the scientific approach in solving the problems that exist for nursing. Agencies in which nurses practice nursing are geared to research being conducted by professionals. Nurses are now being considered professionals. Hence, the time is ripe for nurses to move into the field of research. It is anticipated that administrators, physicians, co-workers in the health care field would indeed acknowledge the need for research in nursing and at the same time lend support in the conducting of various kinds of research. This support will be more readily available if nursing can demonstrate that the findings will be used as a basis for the improvement of nursing care.

The users

The users and the providers of nursing have different characteristics as a result of changes, some planned, some occurring in traditional ways and some occurring by drift. In Saskatchewan, as in much of the Western world, the users of nursing service are a more aged population than previously. In this province the most rapidly enlarging group of people are those over 85 years. This group is predominantly widows, and social surveys show many of them are not financially secure and have had no educational preparation beyond primary or secondary grade school.

It has been acknowledged that the users of nursing — as well as all health services — are better informed than previously about their own health status, options open to them for varying patterns and sources of care, and rights as humans. Because of the varied sources of information, that which consumers accept as being true may conflict with what has been scientifically established as a factual reflection of a given situation.

The expectations of the public regarding their personal health status and the services that should be available to them have changed. Modern means of communication, permitted by satellites, television, etc., have informed people about services available in research centres, teaching hospitals and sophisticated medical centres. The public is aware of the replacement of body parts either by mechanical devices or by donated human organs. People have access to written detailed descriptions of extensive surgical procedures that are being developed to correct human frailties that occur from the aging process. In a world where trauma caused by traffic and industrial accidents is increasingly complex the public is more aware that heroic measures are used to counteract the damage. Surgical procedures, exotic medications, highly technical nutritional management result in many people surviving who otherwise would have died. These persons may be returned to a health status comparable to that prior to the 'health crises' or they may remain disabled in varying degrees and fashions. It is equally true that neonatal units, staffed with highly skilled practitioners using exquisite machinery, are saving newborns who previously would have died. These newborns, just as adults injured or ill, may require varying degrees of care, including home, day, institutional-acute or long term, throughout their life span.

As new techniques and patterns of health care management evolve the human is living longer. This in turn creates a collection of different needs for health care as the individual ages. Care is not restricted to needs related to physical deterioration or alteration but must encompass the psychologically and sociologically engendered needs. Some of these contribute to physical needs developing and vice versa. The social practice of retirement at age 65 may transform a physical well, mentally alert individual who has been contributing on a daily basis to society, into a financially deprived, purposeless creature who requires intensive support services. Not only does this individual require assistance but direction and guidance may be required by the immediate others. The family or surrogate family of any person requiring health care, need and deserve assistance from nurses. The need may relate to understanding what is happening from a disease process point of view, expectations as to prognosis and the anticipated role of the family as the disease process evolves — either with cure, partial return of the patient/client to previous health status or termination of life.

Nurses have been provided with a wide variety of machines which can be used as life support systems, or serve as assessment tools. Theoretically these instruments are seen as time savers whereas in actual practice they are time users. To care for a patient on-line to monitors, analyzers, respirators, vascular pressure readers, intravenous feedings and drainage systems requires not only knowledge of the patient but skill in finding and caring for the individual, psychologically and physically, among the mass of tubes, lines, electrodes, bleepers and buzzers. This is frequently supplemented by the need not only to understand the information the machines are providing, but to understand and perhaps, service the machines themselves.

To determine the best ways — cost effective, efficient and most beneficial to the client/patient — for nursing to keep pace with the changes in health care requires research.

APPROACH TO NURSING RESEARCH WITHIN THE NURSING DEPARTMENT

A nursing research committee

A realistic approach is to structure a nursing research committee, whose ultimate goal would be the improvement of nursing care. The committee could be structured so it will stimulate, foster and facilitate investigation in areas designated as important to nursing and be initially involved in pursuing those things that are totally within the realm of nursing practice. The committee may focus on problem areas: various methods of providing, improving and measuring the quality of nursing care, solving interdepartmental relationships that are generated by the nursing department, investigating questions that nurses have with regard to the kind of care they are providing; developing or comparing different procedures or protocols that are used for delivering nursing care. It is possible that a nursing research committee could be aligned with a medical research committee, whose original focus would be medical concerns which ultimately influence nursing practice. This alliance could result in a joint approach to review items of mutual interest and concern.

When considering research it is important to remember that conclusions based on the knowledge that is available today may in the near or distant future be considered obsolete because of the acquisition of more extensive or different information. In order to develop a nursing research committee that can be functional it is necessary to remember that many of the graduate nurses on staff have gone through a preparation that was relatively authoritarian and did not stimulate the nurses to have inquiring minds. Nurses may have had questioning minds at the time they enrolled in the nursing program. They often found, and quickly, that a questioning behavior was not the type of behavior that obtained the required rewards: passing grades in theory and practice. The practice field was often one in which 'do as you are told and as quickly as possible' received the greatest emphasis. Some of these nurses now, if they question in their minds, do not verbalize the concern; or they assume, when considering the question, that there is no answer and hence there is no point in asking the question.

This means, in selecting a committee, a group of nurses who express an interest in asking questions, in seeking answers to the questions and have a thirst for knowledge is desired. These people then need an opportunity to unlearn their old conditioned habits of accepting the *status quo*, to experience the process of change and to be conditioned to generate, participate in and accept change. To create these changes in persons requires a change agent who is conversant with the change process, knows the target group, is accepted by that group and has the open support of the administrative personnel in the department and the agency. The change agent must have credibility, be capable of making each one of the group feel and be a contributing group member, and be able to develop critical paths associated with specific time periods and accurately assess progress in the areas of concern. The change agent must critically assess the value system adhered to by the committee, the nursing department and agency as it is the basis upon which any research activity will be developed. Following the development of the committee, it then functions as a change agent generating the transition of the nurses from their traditional status to that of a group of professionals, earnestly and eagerly searching for knowledge and methodologies which will enhance the service that can be offered to and provided to clients/patients and their concerned others.

Clinical research implies change

Change: what and where

Change is a six letter word that describes a something, 'a force' that results in a difference. There are synonyms for change but each provides a direction or specifies a sphere in which the change occurs. These synonyms include:

> vary — to differ from time-to-time;
> alter — to bring into line; modify a minor manipulation;
> transform — a difference in form or function;
> convert — to move to a different form or function.

Change takes place in the environment and within resources. Environmental change occurs in the physical and social milieu. Resources which change include the material (structural and financial) and the human [knowledge, attitudes, behavior (skills)].

Ways in which change occurs

There are three major ways by which change is brought about. The first, drift, occurs insidiously and the transformation has occurred and is firmly entrenched before people are aware of the change. This is aptly illustrated by the story of the woodsman's axe. A traveller in Vermont complimented the wood cutter on the beauty and strength of the axe. The woodsman proudly replied 'This is Abe Lincoln's axe — it has had seven new heads and six new handles but it is a wonderful axe'. To the woodcutter it was still the President's axe regardless of the replacement of parts. The change was not realized, indeed not even recognized.

The second form by which change occurs is referred to as traditional. Within this format there are a variety of methods used. Ideas from leaders can be provided to people who associate the idea with qualified persons, and endorse it. As the ideas are adopted change occurs. The elite may exercise power, which is derived from knowledge or from position, and direct that a change occur. The elite may be management or the worker group. If the latter, the power comes from the grass roots where ideas gain force by virtue of the large number of workers involved. Another traditional way to create change is to import a recognized expert in the field. The expert will cite reasons for the change and implement it. People will accept the change because they perceive the expert to be credible and hence worthy of being believed and followed. The last way to implement change traditionally is for a change agent to study the situation, analyse what is happening within the group of people involved and use this information to develop tactics which will persuade people to change. This method has been used by many government agencies who implemented legislation requiring automobile drivers and passengers to wear seat belts (safety harness). Tactics included providing, by newspaper, radio and television, statistical information about accidents, injuries, and deaths occurring when the persons involved were/were not wearing seat belts. Pictures of actual — not simulated — accidents revealed gory details. Following this heightening of awareness of the public of the value of seat belts, legislation which makes wearing seat belts mandatory was enacted. A period of grace before law enforcement officers lay charges against people who did not comply with the legislation was usually allowed. The government will monitor the degree of adherence by the public to the legislation. When the adherence begins to decrease a campaign to increase the awareness of the benefits of wearing seat belts will be initiated. When the incidences of lack of compliance with the law decrease the information campaign will become limited or be withdrawn. When statistics reveal a need for reinforcement the same or similar information campaigns will be introduced.

The most recent form of bringing about change is referred to as 'planned change'. Manufacturers are credited with introducing this concept when products were built to last for a specified period or designed to be replaced when new models appeared. In this form there is an acknowledged intent to move from one point to another at a given time; the people who are affected are to some degree involved in the planned change (purchasers of automobiles are invited to recommend changes that will enhance customer satisfaction with the vehicle); and a group of people are charged with the responsibility of implementing the change. This group becomes a power elite.

Factors which affect research in nursing

Nurses. The members of the profession are influenced by changes within the physical and social environment. As the needs and methods of gratification for individual practitioners change the profession alters. Persons who complete nursing education programs enter the programs with a set of values derived from their home and social environment. The present educational programs differ from those of 5 or more years ago. Newly graduated nurses know more about measurement, are more alert to research findings, have seen results of research applied, may have participated in research projects. These nurses may form a power elite which will precipitate research in nursing practice — not only for the benefit of the client but for the benefit of the nurse or the profession.

The users of the service. Patient/clients have achieved a higher level of general education than in previous decades. They have also acquired a broader knowledge of health from health education classes, and from various features offered by newspapers, magazines, radios and television. In addition people are more cognizant of rights of consumers in all fields including health. This body of knowledge provides consumers with a sound base for developing certain expectations from nursing service. These expectations include the use, by nurses, of proven methodologies based on research.

It must be noted that expectations of consumers may also be based on inaccurate information obtained from a variety of sources. It is essential that nursing possesses soundly researched data from which to refute such claims of a client. This research may be done by nursing or others but nursing must be aware of the data, have access to it and know how to use it.

Education/technology/communication. The volume of information in the physical and social sciences has expanded dramatically. Paralleling this expansion has been the development of technology, which enables replacement of numerous body parts by human transplant and mechanical devices. Machines may temporarily or permanently perform a given body function. The population of the world has been informed of these spectacular activities by an equally spectacular communication system. The communication network gives each of us access to selected events from the entire world in a matter of seconds.

Computer science is now a component of general education. Entertainment relies on computer games. The skills required for these activities belong to the general public. This public is the consumer of nursing service. This public expects nursing to utilize computers as an integral part of service. To incorporate computers effectively into nursing, research is essential. The variety of communication systems provides opportunity for systems of nurse-to-patient, nurse-to-coworkers and nurse-to-nurse information exchange to be developed. The range of possibilities for patient education, monitoring of patient status and measuring and evaluating quality of care are inexhaustible. Research will permit nursing to adopt these systems to enhance nursing care.

Professionalism. As nursing matures it demands a decision making role among its peers. This process will clarify the independent role of nursing, while also acknowledging its interdependent and dependent aspects. The identification, by research, of the role of nursing within the total health field will provide the parameters within which nurses can become independent practitioners — truly professionals. This clarification will require all practicing nurses to become accountable for their own actions. Each one becomes an independent practitioner who, by implied or actual contract with client, agrees to provide a service and accepts the concurrent responsibility and accountability.

Political arena. Nursing, to continue to contribute to the wellbeing of people, must be directly involved in decision making that charts the course for planning, delivery and assessment of health care and the prevention of disabilities and disease. To do so effectively nursing must be involved in and generate research which will clearly indicate the effect (quality outcome) of nursing. This data may then be used in the political arena to further nursing's influence. Research is also needed to identify the mechanics that are most effective in the political field.

Population changes and allied needs. For many years nursing has been concentrated in acute care facilities. Some community nursing, long term care nursing and nursing in physicians' offices has taken place. In recent years the 'special nurse' has not existed. Changes in population size, age distribution and geographic location are major factors currently impacting on the total health/illness scene and especially on nursing. The most rapidly growing age group in Saskatchewan is that of 85 years and over with the greatest number of them female. It is also known that people over 65 years consume the greatest amount of health services provided and that institutionalization is an expensive and dissatisfying method, in many instances, of providing health care for the aged. These realizations are leading to home care programs, supervised residences, day hospitals, and short term periods of care in agencies so families have 'a break' from caring.

Research is required to show the role of nursing in these different kinds of service and whether quality has been affected. Research is also required to substantiate the *role* of nurses as independent practitioners. This latter situation directly relates to the funding, by government agencies, for health care. Research could explore the effectiveness of entry to the health care system through nursing rather than initial contact with the physician.

Ecology/environment. Changes in federal, provincial and municipal regulations governing what people can/cannot do to or in the environment has implications for nursing. Nursing utilizes a great deal of material — disposable and otherwise — in caring for patients/ clients. Nursing is involved in management of chemicals affecting humans and/or the environment. The relationships between environmental contamination and utilization/disposal of chemicals and materials can be demonstrated by research. The results of such research could then be utilized in further research to determine the most effective ways for nursing to manage its impact on the environment.

Dealing with resistance to research

Research as previously noted implies change. The instituting of a research project is itself change, and the perception that the result of the research will necessitate further change may cause nurses in a clinical or other setting to resist the undertaking of the research. The resistance may be overt or covert with the latter being the most difficult to remedy. Steps can be taken which will eliminate this resistance.

One action to be taken early is to import a researcher, who is very knowledgeable of the change process. This person must be given details about the target group and provided with time to get to know the group — as a group and as individuals. This opportunity will also permit the group to develop rapport and establish trust with the expert researcher. It is essential that the expert has full support of management and be able to fraternize with the target group to ensure that each in the group feels a part of it — sharing responsibility for what happens to it. When resistance to research is detected it is essential that the persons in charge listen acutely and observe carefully the comments and actions of the target group. The researcher must assure that everyone understands the research to be undertaken. It is essential that those involved understand their degree of involvement and accept responsibility for their actions. Realizing the worth of their contribution will stimulate them to be accurate in the completion of their tasks.

The researcher must recognize principles of human relations when soliciting support for a project. It is imperative to recognize that different people react differently to the same situation, and that one person may react differently at different points of time to the same situation. The researcher must recognize that values, beliefs, education, experience and social environment govern the way in which one responds to a situation. It is of marked value if one is able to introduce a research project soon after a group has been involved in a positive experience, particularly one which has brought them personal satisfaction or gain. The initial experience will create an open mind toward research and perhaps stimulate them to expect satisfaction from being involved in research.

Tools for research

Tools vary but must be designed to extract the information being sought. Just as a metre stick cannot measure yards a tool designed to assess attitudes cannot measure behavior. To say tools must be specific does not imply that each experimental, historical or descriptive research project must produce new tools. If a tool is available and appropriate, use it. If a unique tool is required for the project then invent or develop it. All tools must be pretested. A questionnaire is a written tool — listing questions which may allow for open ended answers or restrict responses to selection of listed alternatives.

Another form is to select responses from illustrations. A verbal type of questionnaire is the interview. There are varieties of interviews: group/single; guided/open; focused/broad topic. In each of these the questioner may or may not use a set of specific questions to elicit information. In the more controlled, structured interview questions will be asked in the same words, same sequence, same voice quality, etc., to each person being interviewed.

There are a number of paper and pencil tests which can be used. These include matching words with words, definitions or pictures; completing open ended sentences; selecting most appropriate words from a listing of terms; designating statements as true or false. Scales provide a mechanism by which a rater attributes a rank to a certain quality or characteristic. It can be done very simply by an ordinal scale, more distinctly by an interval scale which introduces an equidistant component of measure, and best of all by a ratio scale which identifies a consistent starting point for measurement. Anecdotal records are factual written descriptions of behaviors of persons or activities observed, made at the time and place of observance. The recording does not include interpretation of events or assessment of feelings about the activity. If extended over a specific time period this form of recording may be referred to as a behavioral diary. Checklists are predeveloped documents which, as the name denotes, consist of the listing of a number of items about which an observation will be made. Items on the list may be determined to be present/absent; positive/negative; ranked in order of preference; ranked in degree of effectiveness, etc. The checklist must include specific directives so the observers' responses are directed toward the purpose of the checklist.

Implementing research in the clinical setting

Establishing the environment for research

The nurse administrator must be thoroughly convinced of the value of research. This administrator must endorse the concept that the nursing profession will stagnate, among other professions, from a lack of growth if research is not initiated or if research findings are not used. As previously indicated change is not easy and measurement is often seen as a personal threat. Change requires planning which will result in those being affected by the research developing a sense that they are not being threatened. It is very valuable for them to anticipate that the research will have a positive effect upon them, their patients and the work environment. The nurse administrator who conveys this attitude, together with a demonstrated trust of those involved — researcher, participating staff, resource people, will have a good opportunity to introduce successfully research in the clinical field. It is essential that the clinical staff do not feel that administration is compelling them to change for no apparent reason. A determination by staff that force is being applied may rapidly lead them to muster resistance if only for the sake of resistance. This results in a stalemate and the reason for the proposed change(s) becomes lost amid the techniques of the power struggle. Perhaps the core of successful research is the genuine belief shared by staff, researchers and administrators that the project or projects will result in better care for clients/patients, a more skilled and better informed staff and a working environment which is conducive to both of the preceding.

Topics which can be researched

In the clinical setting, an area which lends itself to descriptive research is the *beliefs of patients about the care* they have received. There are a variety of ways to determine what these beliefs are. Either concurrent or retrospective reviews can be done. The methods may include patient interviews (face-to-face or by telephone) and questionnaires issued on admission, immediately prior to or following discharge. Sociological reviews had indicated that patients will be more apt to express real beliefs after they have left the institution or have separated from the service being assessed. Responding post-separation prevents the nursing personnel or other providers of care labelling the respondent as a 'bad patient' or 'troublemaker'. Once labelled by any component of staff as 'bad' the patient is liable to have to wait longer for the signal light to be answered, be left to last to receive the mail, made to wait for a myriad of reasons. Personnel will not make time to visit or determine if anything is needed.

It is essential that patients/clients understand the purpose of the interview or questionnaire. It is advisable for a contact person to be named so patients can explore further the ramifications of the inquiries. Directives for responding to questions need to be clearly stated in language that is free from jargon and easily understood. Short questions which can be answered yes, no or non-applicable together with opportunity to explain one's choice are desirable. It is also useful to provide opportunity for patients to comment on any aspect of the service provided by the agency — although specific questions may not be included.

As with any research project the sample must meet numerical requirements to ensure the responses are representative of the patients/clients served. Also the tools must be tested so the validity and reliability of each are assured.

The components of care which can be examined include waiting time for admission, feelings about being admitted to and leaving critical care areas (intensive care, cardiac area), special treatment (such as renal dialysis) units and observation units. The clients may be asked to indicate if the personal care they received met their needs and if staff responded to their requests for help, information or services. Information as to whether the individuals received the privacy, rest and sleep which they deemed necessary can be sought.

Discharge planning may be explored. If a sheet of information or a checklist is provided to the patient at the time of discharge one can determine if it is used and why. Also the client may be asked to identify the most useful components of the information as well as to specify things they looked for or anticipated but found were not provided. The consumers may be asked if they received the information they needed about tests and treatments performed in hospital, about caring for themselves at home, when to return to work and when to see their doctor again.

The kind of data collection is of a general nature and must be related to the philosophy and objectives of the nursing department. The tool may be modified to suit a specific area or to obtain data about a certain facet. If seeking information about a specific area, such as a preoperative teaching routine, the questionnaire can be so structured. These questions could include items such as 'Was the visit by the nurse from the operating room helpful?'. Asking for a reason to support the 'yes' or 'no' answer can be done by the word 'why' following the yes/no check boxes. Other items that might be explored include the patients' assessment of the preoperative teaching of deep breathing exercises, coughing routines or crutch walking techniques. These follow-up questions, if applied with sufficient vigor and uniformity, can in the aggregate provide data for measurement of quality of care.

The *environment* can be divided into physical, social, psychological and other components. The topic may be explored as a whole or divided into sections and each examined independently from the others. Another way of dividing this topic for exploration is to determine the way the environment is viewed by the people within selected settings. In the health care delivery system this approach may result in the same milieu being examined from the viewpoint of providers of service and users of service.

One area of environment which has been frequently examined from the patient/client belief is *accommodation* — the single room versus multiple units. This topic was examined frequently in the mid 1960s and early 1970s. There was a general consensus that individuals valued privacy and appreciated the single room. Trends now are leading researchers to re-examine this topic as the clientele using health services have different characteristics than in the previous decades. It is noted that the age is greater, the number of diseases or disabilities being experienced by the individual may be greater, since the client is older the senses especially, hearing and sight, are less acute. It can be anticipated that the responses may vary from those of the earlier decades.

Items which can be examind (Rosso, 1976) from the patients' perspective include whether the individual prefers to be alone or share space with one or more others and their reasons for so feeling. Other information which may be pertinent includes the proximity of home to health care facility and the length of time in hospital. The staff of any department whose work could be influenced by patient placement can be examined. As with the clients, it is essential that health workers be provided with an opportunity to state their rationale for the answers provided. The rationales stated by respondents may serve as the bases for recommendations from a study.

The nursing department assesses self and nurses. Establishing standards. Each department, if it is accepting of its responsibility to the public it serves, needs to audit its performance. This audit can only be done if the standards which the department is committed to meet have been predeveloped. These standards, once developed, need to be tested for validity and reliability as indicators of quality. A study of this type (Willis & O'Shaughnessy, 1980–81) was requested by the Saskatchewan Registered Nurses Association. A total of 95 outcome standards identified in a draft document (Hewitt et al, 1978) were rated by a random sample of directors of nursing in hospitals and directors of care in nursing homes. The findings (Willis & O'Shaughnessy, 1980–81) indicated 39% of the outcomes were not considered valid and reliable. Recommendations suggested revision and subsequent testing of the document. This was done at the end of 1982.

At the Hospital for Sick Children a report (Jenkinson & Weinstein, 1975) describes a project designed to develop tools to measure the quality of care and one to measure the quantity of care. It was then intended to use these tools and their findings when making decisions about staffing. This report describes patient classification systems, the development of the NARvel (Nursing Attention Requirement level) classification with its determiners, validation of NARvel and subsequently the development and validation of the tool for quality measurement title SAVE (Selected Attribute Variable Evaluation). The conclusions included 'there is a positive relationship between quality and NARvel'. It was reiterated throughout the report that the tool was developed for use in a specific hospital within a specific environment. Conclusions also state 'the methods used to develop it are given in detail, and could be reproduced in other settings'. Rosso & O'Shaughnessy (February, 1982) reported findings from two demonstration quality assurance programs in nursing which indicated it is possible to measure the quality of nursing care and identify areas requiring improvement.

At the present time in Saskatchewan the professional nursing association is working closely with the Hospital Systems Study Group to examine the way in which nursing fits into the total patient (and hospital) information systems. The topic is referred to as Nursing Information Systems Saskatchewan (NISS). Hailstone (1982) describes the terms of reference for the Ad Hoc Committee of professional nurses. The committee was charged with identifying the common patient populations for which predeveloped care plans could be prepared and to devise a universal format for such plans. Initially the format was to specify potential problems, related desired outcomes including patient/family teaching within a time frame, nursing actions and standards to be achieved by discharge or separation from the health care facility. Other areas to be explored included concurrent and retrospective audit, patient classification, skill level of staff, recording and reporting. An anticipated outcome of the project was the identification of further action required for the development of nursing information systems. 'An ad hoc committee ... was established in April 1982 to prepare a proposal for testing the NISS forms to ensure that these are reliable and valid and to recommend any revision indicated'. These forms were pilot tested in one large acute care agency in the fall of 1982, followed by testing in one large acute facility in the spring of 1983.

Another approach to identification of standards was undertaken by a group of occupational health nurses in 1979 in Saskatchewan. The group formed a committee which, using as a resource the American Occupational Health Nursing Statement, developed a definition of occupational health nursing. Subsequently, using the format of the Saskatchewan Registered Nurses Association Quality Assurance Program, the committee developed standards and criteria for occupational health nursing. In addition the committee identified care plans for the most common conditions encountered among the clientele served by occupational health nurses. These care plans were accompanied by a form for retrospective nursing audit. Predeveloped nursing care plans included those for fever of unknown origin ('flu'), ambulatory sprain, obesity, GI upset, dysmenorrhea, upper respiratory infection, headache. The Ad Hoc Committee recommended 'that the Occupational Health Nursing Standards and Predeveloped Nursing Care Plans be: 'pretested for clarity, sequence, reliability and validity in two occupational health nursing departments, one in a health care agency and one in an industrial agency' (SRNA, 1980). To date this proposed research project has not been undertaken.

Paralleling the development of the standards for occupational health nursing was a mail survey done primarily to determine salary ranges of the 39 nurses employed in Saskatchewan as occupational health nurses. Also solicited was information about the number of occupational health nursing positions, the clients served (in total and on a daily basis) the number of years each incumbent had been in the position and the kinds of educational programs/courses that incumbents felt they needed in order to fulfill the requirements of their jobs.

Performance expectations. The expectations of employers and employees for self and each other is another topic of research that relates to the clinical field. A 1976 study (SRNA, 1976) sought to determine what were the similarities and differences relative to the expectations of the practitioner and the employers regarding the performance of the new graduate. Also sought was the opinion of the two educational agencies whose programs prepared diploma graduates. This survey examined nursing skills, attitudes and employment characterisitcs. Findings were determined to be useful to: funding bodies for budgetary purposes, employing agencies for planning orientation and inservice programs, nursing administrators for staffing purposes and educational institutions for curriculum planning.

Continuing education. In April 1973 Rosso reported the beliefs of nursing staff of an acute care agency about staff development needs. Programs were deemed essential to retain and improve the quality of performance of an individual. O'Shaughnessy & Conroy (October, 1980) presented a descriptive study of actual participation by nurses in continuing education. The purposes of the study were to determine:

1. SRNA membership participation or non-participation in continuing education programs for 1979.
2. a method for recording SRNA membership participation or non-participation in continuing education programs using electronic data processing methods.

One of the sources of data was the Professional Development Record which nurses seeking renewal of registration are requested, not compelled, to complete. The record provides for each nurse to document study method (attendance at courses, self study, etc.), sponsoring organization (employing agency, continuing nursing education, professional association, union etc.), recording of time to be shown in number of hours and topic. Topics to choose from included Administration, Labour Relations, Clinical Courses, Performance Appraisal with space for other items to be specifically named. Recommendations included making available to planners for continuing nursing education in Saskatchewan the list of suggested topics, that the Professional Development record be done for 1980, 1981, 1982 and 1983 with a pilot test regarding subject areas to be done prior to sending out the 1981 form. These recommendations continue to be acted upon.

A related topic was explored in 1980 with administrative nurses (SRNA, 1980) located in hospitals and nursing homes in Saskatchewan who were surveyed regarding frequency and timing of the provincial meetings of the administrative nurses' special interest group. Educational topics deemed desirable were listed in frequency as cited by respondents. This report was factual and did not include recommendations.

Orientation programs. The professional association in Saskatchewan throughout the last two decades has pressured mainly by resolution, the funding body to make money available so each nurse beginning work in an agency could have a one-month orientation. In addition to this, guidelines for orienting nurses in agencies with a rated bed capacity of 50 or less were provided in 1977. This was supplemented in 1980 by the Suggested Orientation Program Outline for Beginning Nurses in a hospital having 51–250 patients (SRNA, October 1977; August 1980). Questionnaires were issued to the beginning diploma graduates in 1981 and to directors of nursing employing them. This was the third year of an on-going study (O'Shaughnessy, 1982) which is being repeated in 1982.

This writer was involved in an individual agency study to determine 'from analysis of documents completed for and by each graduate of a 1975 basic nursing education program participating in the orientation program, the number who demonstrated achievement within the prescribed time limits, of the skills considered essential to function as a clinical nurse at the Plains Health Centre' (Rosso, 1975). This study showed the total program as being:

Phase I: duration — 2 weeks
            General orientation to the Plains Health Centre and
            orientation to the Nursing Department.
Phase II: duration — 2 weeks
            Orientee performs clinical nursing with a buddy —
            gradually assumes a portion of the buddy's assign-
            ment.
Phase III: duration — 2 weeks
            Orientee functions as a clinical nurse with assistance
            and guidance available from the buddy.
Phase IV: duration — 2 weeks
            Orientee functions independently as a clincial nurse.
Phase V: duration — 4 weeks
            A month of clinical practice following which the
            adaptation of the orientee to the role of clinical nurse
            at the Plains Health Centre would be assessed.

61 nurses participated in the program, each one completing the 12 weeks orientation. This process was available as the health care agency was new and introducing services on a progressive basis hence nurses and other employees came on-stream at intervals. The project also permitted a performance appraisal at the end of the 12-week orientation. This appraisal was a shared endeavor between orientee and the assessor — an assistant director of clinical nursing.

Other studies which relate to nurses and the work environment have been done. These may consider rotas, hours of duty and factors which inhibit nurses returning to the work force. Two of these published by the Saskatchewan Registered Nurses Association and prepared by O'Shaughnessy are Survey of Hours of Work of Registered Nurses in Saskatchewan Hospitals and Nursing Homes for September 1981 and Survey of Inactive Nurses — nurses registered in 1974 who did not register in 1975, nurses registered in 1977 who did not register in 1978 published in 1980. The initial group of 230 nurses was followed for 5 years. Their reasons for becoming inactive remained constant throughout the 5 years i.e., 'family and personal needs' and 'not financially necessary'.

In the follow-up of 258 nurses in the second group the same reasons were cited. One area of difference in the two groups was the priorities given to factors influencing inactive status. Those registered in 1974 but not 1975 listed better wages, less heavy workload, more weekends off, better lines of communication, more progressive atmosphere. The listing by those registered in 1977 but not in 1978 was sequenced as less heavy work-load, more weekends off, better lines of communication and shift work.

SUMMARY

It is apparent from the studies reviewed that the focus in Saskatchewan, at the moment has mainly been on the nurse and the environment. There have been few studies or research projects that have specifically examined the contributions or lack thereof of nursing to the recovery of a person requiring institutional or community care. Some projects have demonstrated that teaching by nurses has been more effective than other disciplines in having clients adhere to weight loss programs or behavior modifications for other reasons.

The possible areas for research in nursing, clinical and non-clinical, are almost inexhaustible. So little of the terrain has been explored that the enterprising individual nurse or non-nurse, can select practically any area of interest and apply historical, descriptive or experimental research methodologies to determine information not currently known. The focus for improving and measuring the quality of care may be on the system of delivery of nursing services, the independent, interdependent and dependent facets of nursing, or the milieu in which nursing is practiced. Featured may be the nurses — what they do, how they do it and why they do it — and the effect of the nursing upon the user of the service. This effect can be explored from the view of nursing service as a maintainer of health along the conception-to-death continuum and as an intervenor to assist the individual to regain optimum health when the body has been assaulted by disease processes or trauma. There is great need for accurate data to be determined about the various fields of practice. From this information educational programs for those already in practice and for those who are selecting nursing as a career/vocation/profession can be constructed which will prepare the provider of nursing to work in the world of today, and have a theory and practice base which will permit and facilitate adaptation of nursing skills to the world and client of tomorrow.

The desirable state of nursing having a sound research base will only develop if each nurse, recognizing that one cannot do everything but that each one can do something, willingly accepts the responsibility for doing that which one can, and subsequently does it. Co-operation will permit the profession to demonstrate that it cares — the core of nursing.

REFERENCES

Hailstone B 1981 Terms of reference Ad Hoc Committee on Nursing Information Systems. Saskatchewan Registered Nurses Association News Bulletin October 1982. 11(6) Regina

Hewitt M, O'Shaughnessy C, Rosso M J 1978 Quality assurance program in nursing practice. Saskatchewan Registered Nurses Association, 2066 Retallack Street, Regina, Canada S4T 2K2

Jenkinson V M, Weinstein E L 1975 The nursing standards project to establish tools for measurement of quality and quantity of nursing at the Hospital for Sick Children Toronto — report for the Ministry of Health of Ontario, Toronto Hospital for Sick Children, Toronto

O'Shaughnessy C September 1976 Performance expectations of diploma nursing graduates — resume of a survey conducted by the Saskatchewan Registered Nurses Association. Saskatchewan Registered Nurses Association, Regina

O'Shaughnessy C March 1980 Survey of administrative nurses. Saskatchewan Registered Nurses Association, Regina

O'Shaughnessy C 1980 Survey of inactive nurses — nurses registered in 1974 who did not register in 1975, nurses registered in 1977 who did not register in 1978. Saskatchewan Registered Nurses Association, Regina

O'Shaughnessy C September 1981 Survey of hours of work of registered nurses in Saskatchewan hospitals and nursing homes. Saskatchewan Registered Nurses Association, Regina

O'Shaughnessy C 1982 Report of the 1981 orientation program for beginning graduates in Saskatchewan hospitals. Saskatchewan Registered Nurses Association, Regina

O'Shaughnessy C, Conroy J October 1980 Results of 1979 study to determine actual participation in continuing education activities in nursing. Saskatchewan Registered Nurses Association, Regina

Rosso M J April 1973 Staff development needs as perceived by the nursing staff Providence Hospital, Moose Jaw, Saskatchewan

Rosso M J 1975 Assessment of the Plains Health Centre twelve-week orientation program for nurses in their initial employment following completion of a basic nursing education program. Plains Health Centre, Regina

Rosso M J 1976 Opinions of patients and employees of the Plains Health Centre regarding single room accommodation. Plains Health Centre, Regina

Rosso M J, O'Shaughnessy C February 1982 Report of the assessment of two demonstration quality assurance programs in nursing. Saskatchewan Registered Nurses Association, Regina

Saskatchewan Registered Nurses Association October 1977 Suggested orientation program outline for registered nurses in a hospital of 50 beds or less. Saskatchewan Registered Nurses Association, Regina

Saskatchewan Registered Nurses Association 1979 Report of the ad hoc committee to develop standards for occupational health nursing. Saskatchewan Registered Nurses Association, Regina

Saskatchewan Registered Nurses Association August 1980 Suggested orientation program outline for beginning nurses in a hospital having 51–250 patients. Saskatchewan Registered Nurses Association, Regina

Willis L, O'Shanughnessy C 1980–81 A project to test nursing department standards for validity and reliability. Saskatchewan Registered Nurses Association, Regina

---

PART TWO

Quality Assurance Movement

Norma M. Lang and Jacqueline F. Clinton

Quality assurance — the idea and its development in the United States

Quality assurance activities in nursing have a long history. Credit for the first documented study in nursing and health care is usually given to Florence Nightingale (1858) for her use of standards to assess care provided to military personnel. For the many decades since Nightingale's study, the assessment and assurance of quality nursing care has remained a priority for nurses throughout the world.

Nursing quality assurance activities in the United States have been numerous over the past two decades. This impressive record of activities can be divided into those of political and organizational, and those of assessment, measurement and research. Underlying these activities are conceptual or philosophical frameworks including definitions. Thus this chapter is divided into a discussion of development of conceptual frameworks, political and organizational actions, and selected empirical research and evaluation studies. A brief discussion of confidentiality issues is also included.

CONCEPTUALIZATION OF QUALITY ASSURANCE IN NURSING

The literal meaning of quality assurance is the assuring of quality nursing care. Operational definitions include a description of a level of quality of nursing care provided and received, measurement of that level of quality and actions taken to ensure that level of nursing care. In contrast, quality assessment is limited to the measurement of quality.

A widely used conceptual model (Fig. 3.1) for quality assurance in nursing is that of the American Nurses' Association (ANA, 1982).

The following is a brief overview of the model (ANA, 1975; ANA & Sutherland, 1982a). The model is circular suggesting that the quality assurance process is continuous. Values are identified, structure, process and outcome standards and criteria are developed

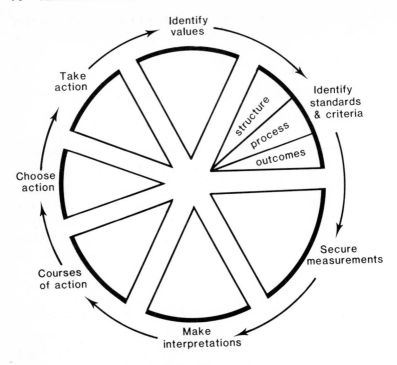

Fig. 3.1   The ANA model for quality assurance in nursing

based upon these values. Tools are developed or selected to measure the criteria. Measurements are taken; the data interpreted to identify strengths and weaknesses of the nursing care. Actions to reinforce or change the nursing care are proposed, selected and taken. The process repeats itself beginning with a reaffirmation of values. Each of these processes are discussed below:

I. Identify values

This first component of the model stresses the need to clarify the social, institutional and individual values that describe quality nursing care. An attempt must be made to arrive at a consensus between the values of the nursing profession, the public, federal government, third party payors, administrators and other health professionals. This clarification is the major challenge to quality assurance as the 1980s become the era of cost containment (Lang, 1982). The ethics of accessibility to a universal standard of care is an example of the problems that need to be addressed. In other words, what level of quality of health care are we willing to support in terms of personal behavior, professional behavior, and public policy?

II. Identify structure, process and outcomes criteria and standards

The three point focus for criteria and standards was proposed by Donabedian (1980). Structure encompasses clearly definable and measurable aspects of human, organizational, environmental and physical resources and standards of practice. Process includes activities, interventions and the sequence of events that constitute caregiving. Outcomes are the consequence to patients' health that can be related to the structure and process of care. Criteria are predetermined elements (based upon identified values) against which aspects of the quality of nursing care can be measured and compared. Standards are the agreed upon level of excellence.

III. Secure measurements needed to determine the degree of attainment of standards and criteria

This component implies that a question about the standards and criteria has been asked. For example, to what degree are the standards being achieved? This is often referred to as the monitoring of standards. What are the known problems in the delivery of nursing care? What additional data are needed? These questions are labeled problem solving. Instruments are then selected or developed and tested to answer the question. Data are collected using a specified method from selected data sources.

IV. Make interpretations about the strengths and weaknesses

Both data that reflect achievement of the criterion as well as the lack of achievement should be examined and interpreted in preparation for making decisions about actions to be taken.

V. Identify possible courses of action

Most reviews of studies yield data that can be used as the basis for decision making and policy making by nurses in clinical practice, administration and research. Decisions may vary from very simple actions to very broad, complex actions requiring policy changes.

VI. and VII. Choose and implement a course of action

Some actions may be taken immediately. Some may require years of planning. Alternative actions must be examined in light of existing resources, organizational structure, knowledge and political climate.

VIII. Start over

The quality assurance cycle begins again. As actions are taken, the progress of nursing practice needs to be reassessed and measured.

FEDERAL INFLUENCE ON QUALITY ASSURANCE DEVELOPMENTS

Federal regulations for the control of quality assurance have had an uneven developmental course. The United States federal government first took action to influence the health care delivery system during the Depression in the 1930s when serious health problems began to emerge in those areas hardest hit by poor economic conditions (Kissick, 1970). The initial issue was not one of quality, but rather access to and financing of care. Thus additional resources and information were provided without specific and restrictive controls. Later, in 1946, another federal program having a great impact on health care was established. Known as the Hill-Burton program, a major infusion of public funds was directed at both the construction of new buildings, and the expansion and modernization of existing ones.

Beginning in 1937 and expanding over the years, federal support has existed for targeted research areas such as heart disease, neurological disease, and blindness. The federal government also became involved in health manpower training and biomedical research during the decades of the 1940s through the 1970s.

Planning and control was not given serious consideration until the federal government took on the enormous responsibility of providing financial coverage for the medical expenses of the elderly and the poor with the passage of Medicare and Medicaid in 1965. Because of the concern for rising costs, manpower shortage and maldistribution, and fragmentation of services, policymakers began to believe that planning and controls were needed.

The first federal program which attempted to tie health spending to more active planning was initiated in 1965 with passage of the

Heart Disease, Cancer, and Stroke Amendments (P.L. 89–239) commonly known as Regional Medical Programs (RMP). The primary mission of RMPs was to promote, through 'Regional co-operative arrangements,' the extension of new medical knowledge about disease treatment from the medical centers into the office of general practitioners and community hospitals. The program was intended to close the gap between health technology discovery and application. Federal controls were not stressed; rather, programs depended on local initiatives.

The second attempt by the federal government to tie health spending to planning and control (Egdahl & Gertman, 1976) was the Comprehensive Health Planning and Public Health Service Amendments (Partnership for Health Act, P.L. 89–749). This law was established to assess health needs and service capabilities, and to establish priorities for future federal and state funding for health programs. Like the RMP, it had limited authority.

The first federal attempt to institute a measure of control (Egdahl & Gertman, 1976) was a requirement included in the Medicare legislation (Social Security Amendments of 1965, P.L. 89–97, Title XVIII) that hospitals and extended care facilities institute utilization review (UR) procedures. These procedures required assurance that the medical services covered under the program was necessary; and that services were provided in the appropriate facility. Still costs continued to rise, thus giving impetus to yet another attempt at federal control, the creation of the Professional Standards Review Organization (PSRO) law designed to overcome the deficiencies in utilization review by changing the organization of the program, broadening the scope of review and ensuring that better use would be made of review determinations. Each PSRO was to develop criteria and standards for the diagnosis and treatment of the cases it reviewed. The goals of the programs were quality assurance and cost containment. However, the challenge of relating these goals remains unachieved.

The most recent federal administration attempt was to do away with the PSRO program. However, the United States Congress passed the Peer Review Improvement Act of 1982 (PRO) with continued funding of local physician organizations to perform review of hospital utilization and quality of care. In addition several proposals for cost containment and competitive health care models, are under consideration. Whether the US direction will be regulation or deregulation or some combination of both was yet to be decided.

THE PRIVATE ROLE IN QUALITY ASSURANCE

The private sector insurance business has sought the co-operation and support of the health professions and hospitals to establish and operate a quality of care and evaluation system for all patients, and has begun to underwrite its share of the cost involved (Pettengill, 1976).

Organizations to review cost of care financed by the federal government (PRO) and by private insurance companies have begun to emerge in the USA during the early 1980s. The future developments in quality assurance will most probably include competitive models, deductibles, co-insurance and other market type mechanisms in an attempt to control the quantity and the quality of care.

VOLUNTARY INFLUENCE ON QUALITY ASSURANCE DEVELOPMENTS

The Joint Commission on Accreditation of Hospitals (JCAH) has been the major voluntary influence on the development of quality assurance activities, especially in hospitals. Since its beginning including its predecessors' organizations, the improvement of the quality of patient care in hospitals has been an explicit goal. Accordingly, in 1952, the JCAH was established to take over from the American College of Surgeons the responsibility of approving hospital programs. The sole purpose of this organization was to encourage the voluntary attainment of uniformly high standards of institutional care. The development of the standards for quality assurance was designed to provide a mechanism for assuring that hospitals evaluated specific segments of patient care and clinical performance. A quality assurance program (QAP) must be comprehensive in nature and focus on patient-care-related problems. The standard requires an adequate review of care across all services, departments, disciplines, and practitioners by nurses, physicians, and other allied health professionals (Egelston, 1980). Beginning in 1983, JCAH will look for evidence of the following when making accreditation decisions: 1. authority and responsibility for quality assurance is assigned in accordance with the description in the written QAP plan; 2. information from quality-related activities is integrated and/or co-ordinated in accordance with the written plan; 3. that patient care problems identified through individual QA functions have been resolved (JCAH, 1982).

Nurses in United States hospitals have been very responsive to the

JCAH quality assurance standards. Numerous workshops (JCAH, 1974) were held aimed at increasing the ability of nurses to conduct nursing audits based on outcome criteria. Although an extensive survey has not been conducted, it is very likely that the majority of United States hospitals have developed numerous sets of outcome criteria and have used these criteria as the basis for retrospective outcome audits of patient records. It can only be hypothesized from a few studies that have been reported in literature that actions taken as a result of these audits have resulted in improved nursing care across the country.

The American Nurses' Association (ANA) has actively promoted the voluntary participation of nurses in quality assurance programs. Following the reorganization of the ANA in 1966, Divisions on Practice were established. Reflecting the concern of the nursing profession to improve the nursing care of patients, the development of standards of nursing practice was considered a major priority for these divisions. In 1973, 1974 and 1976, the generic standards (ANA, 1973c) and standards set forth by each of the divisions were published (ANA, 1973a, 1973b, 1973d, 1974, 1976b). To assist nurses with the implementation of the standards, the ANA has developed programs for peer review, certification and continuing education. Although the ANA has encouraged the use of the standards in the licensing of nurses and the accreditation of nursing educational and service programs, there is no mandate to do so.

The ANA (1975) included licensure, certification, accreditation, peer review as parts of the quality assurance system. Licensure, certification, and accreditation are pertinent to nurses and to the settings in which they practice. These have been described as follows:

Licensure

The process by which an agency of government grants permission to persons to engage in a given profession or occupation by certifying that those licensed have attained the minimal degree of competency necessary to ensure that the public health, safety, and welfare will be reasonably well protected.

Certification

The process by which a nongovernmental agency or association validates, based upon predetermined standards, an individual registered nurse's qualifications, knowledge, and practice in a defined functional or clinical area in nursing; a process in which peers set the education and/or practice requirements that surpass the minimum required for practice.

Accreditation
The process by which an agency or organization evaluates and recognizes an institution or program as meeting certain predetermined criteria or standards. Educational programs, nursing service programs and health care programs are accredited by various organizations.

Peer review
The process by which registered nurses, actively engaged in the practice of nursing, appraise the quality of nursing care in a given situation in accordance with established standards of nursing practice.

It includes the appraisal of nursing care delivered by a group of nurses in a given setting (Nursing Professional Standards Review) and the appraisal of nursing practiced by individual practitioners (Nursing Performance Review).

The efforts of the ANA to advance the state of the art of quality assurance can be traced through several publications. The first conference on quality assurance in nursing was jointly held by ANA and the American Hospital Association (1976). This was followed by leadership workshops throughout the United States to assist nurses with the implementation of standards via a quality assurance program (ANA, 1975). The publication, *Guidelines for Review of Nursing Care at the Local Level* (ANA, 1977), assisted nurses with involvement in the Professional Standards Review Programs in their local areas.

The publication of *Issues in Evaluation and Evaluation Research* resulted from a 1975 ANA conference bringing together evaluators and researchers to make recommendations for future directions (ANA, 1976a). The entire issue of the 1980 March/April issue of Nursing Research summarized the efforts to bring together practicing nurses, evaluators and researchers. In 1982 the ANA published a series of volumes containing specific recommendations for the monitoring of standards and solving of problems via a quality assurance program (ANA & Sutherland, 1982a, 1982b, 1982c, 1982d, 1982e).

Developments in accreditation of educational programs at the national level have been the responsibility of the National League for Nursing (NLN). The NLN also accredits selected organized nursing services. Approval of educational programs leading to the registered nurse credential is done by the state responsible for the licence.

SELECTED EMPIRICAL STUDIES

Increasing attention to health care quality assurance by professional organizations, government agencies, and the private sector served as impetus for the acceleration of research in assessing quality of nursing care.

It is important to mention, especially for those unfamiliar with the research history in nursing quality assurance in the United States, that the majority of recent studies evolved from earlier research done in the late 1960s and early 1970s. During that time, considerable research effort focused on the development of criteria-based, process-oriented tools that are generic across patient populations and health care settings. For example, Wandelt & Ager's (1974) QUAL-PACS instrument is a concurrent evaluation of care received by a patient. The Slater Nursing Competency Scale (Wandelt & Stewart, 1975) is a concurrent measure of individual nurse performance. The Phaneuf Nursing Audit (Phaneuf, 1976) is a retrospective chart audit tool for measuring both patient-focused and nurse-focused indices of quality of care. In a recent article, Phaneuf & Wandelt (1982) pointed out that the eventual goal of these early works is to establish national standards of nursing excellence.

The following discussion includes selected examples of recent research findings published since 1975 on quality of nursing care conducted in the United States. It is organized by clinical focus areas. A comprehensive review of research published since 1975 was conducted by Lang & Clinton and is forthcoming in *Volume II:*

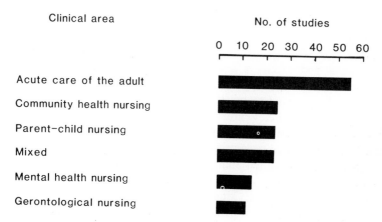

Fig. 3.2 Distribution of research studies on assessing the quality of nursing care by clinical area, 1975–1982

Annual Review of Nursing Research edited by Werley & Fitzpatrick.

Figure 3.2 illustrates the clinical focus of studies published between 1975 and 1982. Nursing care of the hospitalized adult was the clinical area with the largest volume of nursing quality assurance studies, the majority of which focused on instrument development or evaluation of nursing assignment patterns. The younger specialty of gerontological nursing has received less quality assurance research attention compared to other specialties; however, it had the largest proportion of studies (23%) evaluating cost-factors. Evaluation of expanded nursing roles, particularly the performance and impact of the nurse practitioner, was most prevalent in studies of community health nursing (44%) and parent-child nursing (24%).

Nursing care of the hospitalized adult

Since 1975, a substantial amount of pioneering research led to the creation and testing of instruments measuring the various components of nursing quality assurance. One of the most comprehensive tools is the Rush-Medicus instrument developed by Hegyvary et al (1979a, 1979b). It measures structural variables (agency characteristics; nursing unit organization; nursing staff perceptions, attitudes and education levels; and supervisory staff perceptions and expectations) and it also measures the process of care-giving by nurses. Relationships between/among these quality indicators has also been investigated in different patient populations (Hegyvary & Haussmann, 1976). Another comprehensive research effort by Horn & Swain (1978) developed and tested 435 patient outcome criteria amenable to nursing influence. Included are measures of patient health status, attitudes and knowledge of health and health deviations, and self-care agency. Ventura and her associates have done vigorous and extensive reliability and validity testing of quality assurance measures applied to different populations of hospitalized adults (Ventura & Crosby, 1978; Ventura, 1980; Ventura et al, 1980, 1982b; Hageman & Ventura, 1981).

Nursing assignment patterns in the hospital setting (a structural component of quality assurance) have received considerable research attention in the United States. A major comprehensive study of how nursing care is organized and delivered at the hospital unit level was reported by Munson & Clinton (1979). Recognizing that the conventional functional-team-primary descriptors are clear enough to evoke controversy but too vague to permit adequate quality assurance analysis, they developed and tested a tool for measuring the underlying components of any nursing assignment pattern. The tool includes scales for nursing care integration, care management integration, plan-do integration, nursing care continuity, care management continuity, nursing co-ordination, care-cure co-ordination, patient services co-ordination, and intershift co-ordination. The relationships of these underlying 'nursing assignment patterns' elements to other structural variables and patient outcomes were also measured (Munson et al, 1980). Since 1975, a sustained effort has been maintained by nurse researchers to evaluate the impact of various nursing assignment patterns on a wide variety of patient populations and criterion measures of quality assurance (Daeffler, 1975; Jones, 1975; Williams, 1975; Kent, 1977; McCarthy & Schifalacqua, 1978; Eichhorn & Frevert, 1979; Hegedus, 1979; Roberts, 1980; Steckel et al, 1980; Giovannetti, 1980; Bailey & Mayer, 1980; Hamera & O'Connell, 1981; Fairbanks, 1981; Shukla, 1981; Ventura et al, 1982a).

Community health nursing

A major research accomplishment in this area was a national survey of quality assurance activities in public and voluntary community nursing services conducted by Januska et al (1976). Revealed were differences among type and size of agencies as to methods currently in use for assessing quality of care. A summary of future direction for determining quality assurance followed.

Another thrust was evaluation of nurse operated clinics. Kos & Rothberg's (1981) 5 year evaluation of a free-standing nurse clinic in New York documented that quality of care provided by nurses was comparable, sometimes better, than health care provided in more traditional, physician-dominated health systems. However, despite its success in delivering high quality of care as perceived by clients and external auditors, the clinic closed because of reimbursement policies and the thwarting of attempts to interface with other provider systems. Another major longitudinal study on nurse operated clinics (Runyan et al, 1980) documented that nurses delivered high quality, low cost care to populations with chronic conditions resulting in a decrease in hospitalization rates by 47% and a dramatic decline in life-threatening complications by 69%. The introduction of nurse practitioners and a video communication network for physical consultation in a prison system was found to increase quality of care and decrease health care costs (Sanders et al, 1976).

Parent-child nursing

The development of reliable and valid indicators that can be utilized in quality assurance has been one research focus in the area of parent-child nursing. For example, Weinstein's (1976) Selected Attribute Variable Evaluator was developed for evaluating standards of nursing practice given to hospitalized children based on Wandelt & Ager's QUALPACS (1974). Measurement of parenting behavior and coping was tested by Blair et al (1978) and Meleis & Swendsen (1977). Johnson (1977) tested a tool for measuring interactional deprivation between mothers and their premature infants. Barnard & Eyers (1977) developed and tested instruments for nursing child assessment and early maternal-child interaction.

Another quality assurance related research trend has been longitudinal evaluation of new models of comprehensive, multidisciplinary health promotion for mothers and/or children. For example, an interdisciplinary health program established in public high schools was shown to improve significantly the health status of pregnant adolescents and their infants (Berg et al, 1979). Implementation of a multidisciplinary health assessment program in local schools resulted in improved health among preschool children at lower costs than traditional physician care (Robertson et al, 1976). In two major urban settings, comprehensive well-child care involving nurses was found superior in quality to that provided by physicians alone (De-Maio, 1981; Stewart, 1981). Diers (1981) evaluation of nurse-midwifery as a system of care indicated that, compared to medically-oriented services, nurse-midwifery care was less intrusive and more personalized and resulted in higher birthweights and significantly lower incidences of prematurity and neonatal deaths. Lubic (1981) also documented the health benefits and economic savings of nurse-midwifery services.

Mental health nursing

Much of the recent quality assurance research in the area of mental health nursing has been directed to the development and testing of outcome criteria. For example, Newman (1982) demonstrated the usefulness of the Functional Baseline Scale for measuring patient progress on four subscales: emotional functioning, social functioning, task orientation and skill capacity. Guy & Moore (1982) tested the Goal Attainment Scale for use in monitoring patient responses to treatment in psychiatric hospitals.

A second major focus of quality assurance research in mental health nursing has been the evaluation of different nursing models of service delivery. Examples of these are Ginsberg & Marks (1977) impressive documentation on the impact of nurse psychotherapists on the quality of life of both patients and their families and Vincent & Price's (1977) documentation of the positive impact of public health nurse visits to patients discharged from residential treatment facilities.

Gerontological nursing

Development of standards and outcome criteria were prevalent research concerns in gerontological nursing. For example, utilization reviews of numerous nursing homes led to the creation of a Level of Care Scale and subsequent reduction in workload demands for skilled professional judgement (Kane et al, 1981). Self-care outcome criteria for residents in extended care facilities were tested by Howe et al (1982).

Another focus in gerontological quality assurance research was evaluation of multidisciplinary, comprehensive health programs which resulted in improved health status of the elderly and lower hospital admission rates (Polliack & Shavitt, 1977; Sullivan & Armignacco, 1979).

CONFIDENTIALITY

Confidentiality of data is an issue frequently associated with quality assurance. The availability and use of evaluative data about a recipient of care, and/or groups of recipients has always been a concern to health professionals. More recently the concern has also extended to confidentiality of health professional provider data.

The increasing use of computerized data about individuals, including health data, has generated enormous concern for the potential risks to right to privacy assured by the US Constitution. Several principles are suggested by Epstein (1976: p.168–169) to help answer questions about due process and the right to privacy. These principles include:

1. The right of the public to know that a data system exists;
2. The right of an individual to know that a data system includes information on him or her;
3. The right of an individual to have access to information main-

tained on him or her in such a system; and

4. The right of an individual to challenge the quality and accuracy of the information maintained on him or her.

Right to privacy questions include:

1. What information should be collected about an individual through compulsory means?

2. Should information be linked and matched with other information and, if so, on what terms?

3. What judgements should be made on the basis of this information?

The concern is widespread as is reflected by an English author (Pheby, 1982) who writes of the great anxiety which resulted from the growth of routine collection of data relating to individual patients for statistical purposes. Obligations for the reporting of these data imperil the confidentiality of the health professional/patient relationship.

Most quality assurance studies can be conducted without recording of individual names and are reported in terms of aggregate data. Review of care provided to an individual identified patient requires constant vigilance to ensure the protection of patient identity. The U.S. judicial and legislative systems are both involved in determinations that will set policies about use of data about a citizen.

Confidentiality for provider data, whether an individual, group, or institution is also a complex issue. Two conflicting viewpoints are well described by Epstein (1976). Proponents of full disclosure argue that provider patterns of practice and profiles of care should be public information if the consumer of health services is to make rational choices of health care and if a higher level of quality of care can be forced by public attention. Arguing against disclosure is the relatively poor state of the art regarding the definition and the measurement of quality of care; also the concern that data submitted may not be as candid if publication is anticipated. Finally there is also concern for privacy rights of individual providers.

The concern for confidentiality for recipient and provider data is also a major barrier to the advancement of evaluation research. Evaluation or assessments are frequently carried out for a specific institution or practice with the promise of confidentiality. The major purposes of these evaluations is to provide data for clinical and administrative decision making. It is very difficult to get comparative data from which generalizations about quality can be made. The problem summarized by Prescott (1978) is that results of such short range focus and reliance on single program or practice evaluation have led to a large body of evaluation research with little or no generalizability.

NURSES AND PUBLIC POLICY RELATED TO QUALITY ASSURANCE

Participation of nurses in federal and voluntary quality assurance activities are serious and numerous. For the most part, federal and voluntary mandates have specifically included nursing practice. Within the nursing profession, and without government or voluntary mandates, nurses have developed methods to review the quality of care. Because of the large number of nursing personnel (over 2 500 000) in the United States (American Medical Association, 1982), nurses are in a strong position to influence the health care system. Their force is not only in numbers, but in scope of influence. They serve clients 24 hours a day, 7 days a week and are in a strategic position to identify patient care needs and improve patient care.

Nursing has been described as the slumbering giant of the health professions (Aiken, 1981). Indeed, nurses are more numerous than all the other health professional groups combined. Although their influence on national health care policy in the past has not been commensurate with their numbers, important changes have occurred in the past two decades, both within nursing and in society, that may dramatically affect the role of nursing in health care in the 1980s.

Nurses also have an extensive research base that supports nursing practice. The challenge of putting numbers, scope of influence and research findings together with the aim of influencing public policy has only begun to be accomplished.

Aiken (1981: p.15) summarizes the need for continual attention to evaluation and research:

> In situations where objective, rigorous information can be obtained on the costs and benefits of nursing care, the long term strategy that appears to have the greatest payoff is the evaluation and demonstration of nursing care. It must be demonstrated that nursing care can bring about important changes iin the health and well-being of individuals and that resources can be reallocated in ways that are more related to patient outcomes.

SUMMARY

The American Nurses' Association's conceptual model for quality assurance was presented. The model included widely accepted components in quality assurance programs. An overview of the development of public policy regarding the assessment and the assurance of quality of nursing care included such programs as these: Joint Commission on Accreditation of Hospitals, Professional Standards Review Organizations, the American Nurses' Association, National League for Nursing. Empirical work in the assessment of the quality of nursing care was highlighted. Finally, selected legal dimensions such as confidentiality and right to privacy were explored.

Assessment and assurance of quality nursing care remains a challenge to public policy makers, practicing nurses, and researchers. Although the record of quality assurance activities in nursing in the United States is impressive, there is much work to be done in the area of evaluation research including cost-benefit analysis and in utilizing the findings of such research in the arena of public policy. The ultimate goal is to improve the quality of health and nursing care that can be assured to the American public at an affordable cost.

Acknowledgement

The authors acknowledge the assistance of the following graduate assistants: Marian Hein, Liz Weindorfer, Jane Maskrey, Sharon Burr, and Mary Kay O'Brien.

REFERENCES

Aiken L (ed) 1981 Health policy and nursing practice. McGraw-Hill, New York

American Medical Association 1982 The American health care system. The American Medical Association, Chicago, Illinois

American Nurses' Association 1973a Standards of community health nursing practice. American Nurses' Association, Kansas City, Missouri

American Nurses' Association 1973b Standards of maternal-child health nursing practice. American Nurses' Association, Kansas City, Missouri

American Nurses' Association 1973c Standards of nursing practice. American Nurses' Association, Kansas City, Missouri

American Nurses' Association 1973d Standards of psychiatric-mental health nursing practice. American Nurses' Association, Kansas City, Missouri

American Nurses' Association 1974 Standards of medical-surgical nursing practice. American Nurses' Association, Kansas City, Missouri

American Nurses' Association 1975 A plan for implementation of the standards of nursing practice. American Nurses' Association, Kansas City, Missouri

American Nurses' Association 1976a Issues in evaluation research. American Nurses' Association, Kansas City, Missouri

American Nurses' Association 1976b Standards of gerontological nursing practice. American Nurses' Association, Kansas City, Missouri

American Nurses' Association 1977 Guidelines for review of nursing care at the local level. American Nurses' Association, Kansas City, Missouri

American Nurses' Association, American Hospital Association 1976 Quality assurance for nursing care. American Nurses' Association, Kansas City, Missouri

American Nurses' Association, Sutherland Learning Associates Inc. 1982a Professional nurses' role in quality assurance. Nursing quality assurance management learning system, vol. 1. American Nurses' Association, Kansas City, Missouri

American Nurses' Association, Sutherland Learning Associates Inc. 1982b Workbook for QA committee members: General practice in acute care hospitals. Nursing quality assurance management learning system, vol. 2. American Nurses' Association, Kansas City, Missouri

American Nurses' Association, Sutherland Learning Associates Inc. 1982c Workbook for QA committee members: Community health agencies. Nursing quality assurance management learning system, vol. 3. American Nurses' Association, Kansas City, Missouri

American Nurses' Association, Sutherland Learning Associates Inc. 1982d Workbook for QA committee members: Long term care facilities. Nursing quality assurance management learning system, vol. 4. American Nurses' Association, Kansas City, Missouri

American Nurses' Association, Sutherland Learning Associates Inc. 1982e Guide for nursing QA coordinators and administrators. Nursing quality assurance management learning system, vol. 5. American Nurses' Association, Kansas City, Missouri

Bailey K, Mayer G 1980 Evaluation of the implementation of primary nursing. Nursing Dimensions 7(4): 82–84

Barnard K E, Eyers S 1977 Nursing child assessment: The first twelve months of life. Division of Nursing, Bureau of Health Resources Administration, U.S. Department of Health, Education and Welfare, Washington, D.C.

Berg M, Taylor B, Edwards L E, Hakanson E Y 1979 Prenatal care for pregnant adolescents in a public high school. Journal of School Health 49(1): 32–35

Blair E, Hauf B, Loveridge C, Murphy J, Roth M 1978 Instrument development: Measuring quality outcomes in ambulatory maternal-child nursing. Nursing Administration Quarterly: Research Impact on Patient Care 2(4): 81–93

Clermont H 1982 Confidentiality. Worcester Medical News January/February: 5–7

Daeffler R J 1975 Patients' perception of care under team and primary nursing. Journal of Nursing Administration 5(3): 20–26

DeMaio D J 1981 Health services for children: A descriptive analysis of an urban program. In: Aiken L (ed) Health policy and nursing practice. McGraw-Hill, New York, p158–182

Diers D K 1981 Nurse-midwifery as a system of care: Provider process and patient outcome. In: Aiken L (ed) Health Policy and nursing practice. McGraw-Hill, New York, p 73–89

Donabedian A 1980 The definition of quality and approaches to its assessment. Explorations in quality assessment and monitoring, vol. 1. Health Administration Press, Ann Arbor, Michigan

Egdahl R H, Gertman P M 1976 Quality assurance in health care. Aspen Systems Corporation, Germantown, Maryland

Egelston E M 1980 New JCAH standard on quality assurance. Nursing Research 29(2): 113–114

Eichhorn M L, Frevert E I 1979 Evaluation of a primary nursing system using the quality patient care scale. Journal of Nursing Administration 9(10): 11–15

Epstein S 1976 Confidentiality and quality of care. In: Egdahl R H, Gertman P M (eds) Quality assurance in health care. Aspen Systems Corporation, Germantown, Maryland, p 167–176

Fairbanks J E 1981 Primary nursing: more data. Nursing Administration Quarterly

5(3): 51–62

Ginsberg G, Marks I 1977 Costs and benefits of behavioral psychotherapy: A pilot study of neurotics treated by nurse-therapists. Psychological Medicine 7: 685–700

Giovannetti P A 1980 Comparison of team and primary nursing care system. Nursing Dimensions 7(4): 96–100

Guy M E, Moore L S 1982 The goal attainment scale for psychiatric inpatients: Development and use of a quality assurance tool. Quality Review Bulletin June: 19–29

Hageman P T, Ventura M R 1981 Utilizing patient outcome criteria to measure the effects of a medication teaching regimen. Western Journal of Nursing Research 3(1): 25–33

Hamera E, O'Connell K A 1981 Patient-centered variables in primary and team nursing. Research in Nursing and Health 4: 183–192

Hegedus K S 1979 A patient outcome criterion measure: Volicer Hospital Stress Rating Scale. Supervisor Nurse 10(1): 40–45

Hegyvary S T, Haussmann R K D 1976 Monitoring nursing care quality. Journal of Nursing Administration 6(9): 3–9

Hegyvary S T, Haussmann R K D, Kronman B 1979a Monitoring quality of nursing care, part four: The nursing process framework in four speciality areas. Medicus Systems Corporation, Chicago, Illinois (National Technical Information Service No. HRP-0900639/6)

Hegyvary S T, Haussmann R K D, Kronman B, Burke M 1979b User's manual for Rush-Medicus nursing process monitoring methodology. Medicus Systems Corporation, Chicago, Illinois (National Technical Information Service No. HRP-0900638/8)

Horn B J, Swain M A 1978 Criterion measures of nursing care quality. National Center for Health Research, Hyattsville, Missouri (National Technical Information Service No. PB-287 449/3GA)

Howe M J, Coulton M R, Almon G M, Sandrick K M 1982 Use of scaled outcome criteria for a target patient population. Quality Review Bulletin Spring: 39–45

Januska C, Engle J, Wood J 1976 Status of quality assurance in public health nursing. American Public Health Association December: 1–39

Johnson S H 1977 Data-gathering tool on interactional deprivation of mother and premature infant. Communicating Nursing Research 9 WICHE: 102–110

Joint Commission on Accreditation of Hospitals 1974 The PEP primer: Performance evaluation procedure for auditing and improving patient care. Joint Commission on Accreditation, Chicago, Illinois

Joint Commission on Accreditation of Hospitals 1982 JCAH Perspectives. Joint Commission on Accreditation of Hospitals 2: 1

Jones K 1975 Study documents effects of primary nursing on renal transplant patients. Hospitals 49(24): 85–89

Kane R L, Rubenstein L Z, Brook R H, Van Ryzin J, Masthay P, Schoenrich E, Harrell B 1981 Utilization review in nursing homes: Making implicit level-of-care judgements explicit. Medical Care 19(1): 3–13

Kent L A 1977 Outcomes of a comparative study of primary, team, and case methods of nursing care delivery in terms of quality of patient care and staff satisfaction in six western region hospitals. Unpublished study of the Primary Nursing Research Group, Regional Program for Nursing Research Development, Western Interstate Commission for Higher Education, Boulder, Colorado

Kissick W L 1970 Health policy directions for the 1970s. New England Journal of Medicine 282(24): 1343–1354

Kos B A, Rothberg J S 1981 Evaluation of a free standing nurse clinic. In: Aiken L (ed) Health policy and nursing practice. McGraw-Hill, New York, p 19–42

Lang N 1982 Introduction. Quality Review Bulletin/Special Edition on Nursing Review: Criteria for Evaluation and Analysis of Patient Care Spring: 2

Lang N, Clinton J 1984 Assessment of quality of nursing care. In: Werley H,

Fitzpatrick J (eds) Annual review of nursing research: vol. II. Springer, New York (in press)

Lubic R W 1981 Evaluation of an out-of-hospital maternity center for low risk patients. In: Aiken L (ed) Health Policy and nursing practice. McGraw-Hill, New York, p 90–116

McCarthy D, Schifalacqua M M 1978 Primary nursing: Its implementation and six month outcome. Journal of Nursing Administration 8(5): 29–32

Meleis A I, Swendsen L 1977 Does nursing intervention make a difference? A test of role supplementation. Communicating Nursing Research 8 WICHE: 308–324

Munson F, Clinton J 1979 Defining nursing assignment patterns. Nursing Research 28(4): 243–249

Munson F, Beckmann J S, Clinton J, Kever C, Simms L 1980 Nursing assignment pattern user's manual. Health Services Administration Press, Ann Arbor, Michigan

Newman F 1982 Outcome evaluation and quality assurance in mental health. Quality Review Bulletin April: 27–31

Nightingale F 1858 Notes on matters affecting the health, efficiency, and hospital administration of the British army. Harrison and Sons, London

Pettengill D W 1976 The private role of financing of quality assurance systems. In: Egdahl R H, Gertman P M (eds) Quality assurance in health care. Aspen Systems Corporation, Germantown, Maryland, p 271–277

Phaneuf M C 1976 The nursing audit and self-regulation in nursing practice, 2nd edn. Appleton-Century-Crofts, New York

Phaneuf M C, Wandelt M A 1982 Three methods of process-oriented nursing evaluation. Quality Review Bulletin/Special Edition on Nursing Review: Criteria for Evaluation and Analysis of Patient Care Spring: 32–38

Pheby D F H 1982 Changing practice on confidentiality: A cause for concern. Journal of Medical Ethics 8: 12–24

Polliack M R, Shavitt N 1977 Utilization of hospital inpatient services by the elderly. Journal of American Geriatric Society 25(8): 364–367

Prescott P A 1978 Evaluation research: Issues in evaluation of nursing programs. Nursing Administration Quarterly 2(4): 63–80

Roberts L E 1980 Primary nursing? Do patients like it — Are nurses satisfied — Does it cost more? Canadian Nurse 76(11): 20–23

Robertson L H, McDonnell K, Scott J 1976 Nursing health assessment of preschool children in Perth County. Canadian Journal of Public Health 67(4): 300–304

Runyan J W, VanderZwaag R V, Joyner M B, Miller S T 1980 The Memphis diabetes continuing care program. Diabetes Care 3(2): 382–386

Sanders J H, Sasmor L, Natiello T A 1976 An evaluation of the impact of communication technology and improved medical protocol on health care delivery in penal institutions. Volume IV. Telemedicine system manuals. Westinghouse Electric Corporation, Health Systems Division, Pittsburgh, Pennsylvania (National Technical Information Service No. PB80–105893)

Shukla R K 1981 Structure vs. people in primary nursing: An inquiry. Nursing Research 30(4): 236–241

Steckel S, Barnfather J, Owens M 1980 Implementing primary nursing within a research design. Nursing Dimensions 7(4): 78–81

Stewart R F 1981 Evolution and implementation of a comprehensive health care center: Nursing in a pivotal role. In: Aiken L (ed) Health policy and nursing practice. McGraw-Hill, New York, p 43–63

Sullivan J A, Armignacco F 1979 Effectiveness of a comprehensive health program for the well-elderly: By community health nurses. Nursing Research 28(2): 70–75

Ventura M R 1980 Correlation between the quality patient care scale and the Phaneuf audit. International Journal of Nursing Studies 17(3): 155–162

Ventura M, Crosby F 1978 Preparing the nurse observer to use the quality patient care scale: A modular approach. Journal of Continuing Education in Nursing 9(6): 37–40

Ventura M R, Fox R N, Corley M C, Mercurio S M 1982a A patient satisfaction measure as a criterion to evaluate primary nursing. Nursing Research 31(4): 226–230

Ventura M R, Hageman P T, Slakter M J, Fox R N 1982b Correlations of two quality of nursing care measures. Research in Nursing and Health 5(1): 37–43

Vincent P, Price J R 1977 Evaluation of a VNA mental health project. Nursing Research 26(5): 361–367

Wandelt M A, Ager J W 1974 Quality patient care scale. Appleton-Century-Crofts, New York

Wandelt M A, Stewart D S 1975 Slater nursing competencies rating scale. Appleton-Century-Crofts, New York

Weinstein E 1976 Developing a measure of the quality of nursing care. Journal of Nursing Administration 6(6): 1–3

Williams L B 1975 Evaluation of nursing care: A primary nursing project. Supervisor Nurse 6(1): 32+

Quality assurance in nursing: the Australian experience

The improvement of nursing practice has always been the foremost objective of organised nursing in Australia. Although the concepts of modern quality assurance programs are new to many Australian nurses quality has always been a major concern. At present there is an upsurge of interest in the nature of the implied social contract between society and nurses and a steadily increasing awareness of all that is inherent in the claim of accountability. Many nurses are actively seeking to assure quality.

Quality assurance embraces the activities of problem identification and problem solution. The objective is to create a nursing care system that is responsible to the needs of Australian citizens and is efficient and effective. A nursing care system is made up of all the aspects of nursing in which nurses are engaged in whatever setting. Nursing practice includes clinical practice, nursing education, administration, and research. Each of these aspects of nursing practice is dependent upon, and interactive with, the other. Effectiveness and efficiency in each is essential for the proper function of the system as a whole.

Australian nurses recognise the need for unity and an organised approach to quality assurance. Flexibility, adaptability, and a capacity to accommodate advances in technology, nursing knowledge, and evaluation technique are considered essential in a quality assurance program. A national approach seems to offer the best opportunity to capitalise on the innovativeness and creativity that is evident in large measure among Australian nurses. A national approach should increase unity and also avoid fragmentation and duplication of effort.

In this chapter an atttempt is made to outline the major factors that affect nursing in Australia, to report what has been done and the reasons for these actions taken to direct attention to formal methods of assuring quality. No attempt has been made to discuss the theoretical concepts that underlie quality assurance programs. Nor have the advantages and disadvantages of particular systems been discussed except to outline the reasons, as far as is practicable, of the course that is being pursued in Australia.

The effects of the programs being undertaken have not been evaluated, and there is no formal evidence of their impact. Nevertheless, giant steps have been taken to assist nurses to set standards, systematically review nursing practice, identify problems, and take actions to achieve and maintain quality.

FACTORS AFFECTING NURSING IN AUSTRALIA

The national nurses' organisation

The Royal Australian Nursing Federation (RANF) is both the professional and industrial national association for nurses. RANF is a federation of branches; one in each of the eight States and Territories. Registered and enrolled nurses are accepted into membership by the branches. Each branch is directed by a council elected by the members. The branch council deals with the professional and industrial matters that affect nurses and nursing practice in its State. Matters that affect nurses or nursing practice across the borders of a State are dealt with by the total membership acting through the Federal Council.

The Federal Council is made up of the president of each branch and a federal councillor elected by members in each branch. Members of Federal Council, by collegiate election, select the federal executive. The Federal Secretary, elected by the total membership every 4 years, is the chief executive officer of the Federation.

Socio-economic welfare of nurses

Matters affecting the socio-economic welfare of nurses are dealt with under the Australian Conciliation and Arbitration Act, 1904. The Conciliation and Arbitration Commission makes awards, binding on employers and nurses, which define salaries, hours, leave, sickness entitlements, and other conditions.

Each RANF branch deals through the Conciliation and Arbitration Commission in its State/Territory so that there are awards for each State and Territory. The Federation deals with the Commonwealth Conciliation and Arbitration Commission to obtain awards for nurses employed in areas under Federal Government jurisdiction. This system results in some 128 separate awards affecting nurses in Australia. There is lack of uniformity in salaries and working conditions.

Licensure of nurses in Australia

License to practise nursing is granted by the nurse registering authority in each State or Territory. Authority for licensure arises from the State/Territory Nurses Act, consequently requirements and conditions differ in each jurisdiction. Nurses licensed in one State who wish to practise in another must obtain registration in that State.

Funding and decision making in the health care service

Health care in Australia is the responsibility of the States/Territory governments and the Federal Government. Funds for health care are made available to the States by the Federal Government. Consequently allocation of funds to health care agencies is in the control of the State/Territory health authority. However, the Federal Government controls social welfare payments such as pensions and unemployment benefits, and also the matters related to veterans' health.

Escalating costs of health care have resulted in numerous cost cutting measures. In some instances, the number of nurses employed is being decreased. It seems that administrators feel justified in decreasing the number of nurses as long as the work gets done. Perhaps under the erroneous assumption that student labour is cheaper, students of nursing provide the major part of the nursing care in hospitals.

Nursing practice in Australia

Although there are increasing numbers of nurses working in community centres or providing domiciliary care, most nurses work in hospitals or similar institutions. Only a few nurses are working as independent practitioners.

There is a depressingly large number of nurses, particularly in remote and rural areas, who have not had the opportunity for continuing education. Consequently, there has been little stimulus to keep abreast of national and international issues and trends in nursing. It is increasingly difficult for nurses in these circumstances to see beyond the problems directly confronting them in the work place and to draw upon the experience of others to provide a fresh perspective.

In many instances, nurses continue to undertake duties, tasks, and responsibilities unrelated to direct patient care. Nurses have frequently stated that there is no choice for them. A commonly heard question is 'If we don't do it, who will?' and a frequent statement is 'If we don't do it, the patients will suffer'. Administrators have failed to provide the appropriate category of staff to undertake support services. Work, unrelated to direct patient care, has come to be accepted by many administrators and nurses as part of the nurse's role. With this view it is difficult for nurses to see ways of altering the situation unless more nurses are employed.

Arising from the undertaking of non-nursing work on a regular basis is a tendency for a nurse to see herself, and to be seen by others, as a 'jack of all trades'. The true work of the nurse and its worth are obscured. This has allowed decision makers to view nursing as requiring minimal skills. Despite the fact that RANF has articulated policies relating to the skills required by nurses, administrators employ workers other than nurses to provide so-called basic nursing care. In a great many nursing settings such as nursing homes, psychiatric hospitals, homes for the profoundly physically and intellectually handicapped, and rural and remote centres, workers without formal nursing preparation are providing care.

Thus in many institutions the care tends to be custodial in nature and unrelated to the particular needs of individual patients. There is a tendency to view medical orders as the primary determinant of nursing care. This dependence on medical practitioners perpetuates the view that nursing is indistinguishable in many aspects from medical practice and is subservient to it. Such views increase the opportunity for medical practitioners to affect high level decisions about nursing practice and nurses. Adherence to the view that nursing care is initiated by the medical practitioner makes it difficult for nurses to focus on an holistic approach to the client. The effect is a lack of attention to the preventative and educational aspects of nursing practice and an emphasis on the curative aspects of the medical care regime.

Use of unskilled workers and students of nursing to provide a large proportion of care in Australian institutions has resulted in the need for a large number of supervisory nurses. Inevitably this means that the most expert nurses are promoted away from direct patient care. Nursing organisational charts show a pronounced hierarchical structure. As a consequence direct care givers feel unable to affect decision making which they perceive as taking place among the higher echelons.

Nurses in these situations feel powerless to effect change. Often they feel it useless to make suggestions for improvement. Some nurses express the view that the responsibility for change rests with administrators and executive officers. These nurses do not recognise that the functioning of various units and the activities of the individual nurses within the units determine the quality of care provided. Individual and collective responsibility to control nursing care practices is not universally accepted or exercised.

Education for quality in Australia

Despite increasing calls for reform within the health care system and for evaluative research, little effort has been made to educate health workers about the concepts of quality assurance. One organisation, the Peer Review Resource Centre (PRRC), founded jointly by the Australian Medical Association and the Australian Council on Hospital Standards, a body concerned with accreditation of hospitals, attracts Federal Government funds for education of health care workers. To date, most of the activities of the PRRC have centred around collecting information about what is being done by various disciplines in hospitals and other institutions, and development of methods to review medical care and utilisation of medical services.

The lack of training in the concepts of quality assurance and a persistent tendency to view lack of knowledge as the chief, if not only, reason for unsatisfactory performance has resulted in concentration on education as the only possible remedy of poor performance. The RANF recognises the need to educate nurses about quality assurance and to assist them to consider all the environmental factors such as division of labour, delegation of tasks, leadership style, lack of clearly stated standards, and lack of feedback which may affect performance.

The national quality assurance program being developed and implemented by the RANF is designed to help nurses view the nursing system in its entirety so that the focus is on all the aspects which impinge upon nursing practice. Such an approach is considered fundamental to effective problem identification and problem solution for nursing.

STRATEGIES FOR OVERCOMING PROBLEMS CONFRONTING NURSING

Recognising the problems resulting from the diverse controls and constraints on nursing practice, the RANF Federal Council in the early 1970s formed a working party to examine the issues. The working party was to make recommendations which would allow nurses to confront the problems and plan for the future on a national basis. The working party produced two documents of great significance to Australian nurses.

The first, entitled *Goals in Nursing Education* (Donaghue, 1975) sets forth recommendations for systematically moving nursing education from the traditional hospital-based school, apprenticeship-type training to the mainstream of tertiary education. Although not all nurses agree that it is necessary or desirable to change the education system, a great deal of time, energy, and money have been spent in seeking this goal.

In Australia, at present, some 1200 places are available for nurses in Colleges of Advanced Education. There are courses leading to registration as a nurse. In a number of schools or departments of nursing within the colleges, courses for registered nurses leading to the Bachelor of Applied Science (Nursing) with major in various aspects of clinical practice, education or administration and research are conducted.

Since 1979 the Australian-American Education Foundation has offered long and short term W.K. Kellogg Fellowships. Under this program, many Australian nurses have studied nursing in America. The impact of these graduates of Australian and American programs on nursing in Australia has not been determined. However, the impression is that these nurses are positively influencing nursing practice in their work place.

The second document prepared by the working party was entitled *Goals in Nursing Practice* (Donaghue, 1977). This work provided an excellent review of the English language literature. The central theme was the need for planned programs to introduce a problem solving mode in all aspects of nursing. This was seen as necessary to allow collection of the data required to plan future directions and identify strategies required for nurses to achieve the objectives of quality practice and care.

The working party recognised that nurses had not always had sufficient impact on decisions about health care. The lack of consultation with nurses and the dominance of medical opinion was considered to result in imbalanced decisions about health services. A need for the RANF, speaking for organised nursing in Australia, to increase opportunities for nursing expertise to be employed by policy making bodies was considered urgent. Data based on nursing research and quality assurance evaluation, rather than opinion and intuition, was seen to be essential to achieving the goals.

A national approach

In 1979, the Federal Council endorsed a national, unified approach to implementation of the problem solving mode. The RANF branches, acting through Federal Council, agreed to support the program and the national quality assurance program was launched.

A nursing officer responsible through the Federal Secretary to Federal Council was appointed. This officer was given responsibility for devising a plan to achieve a unified, co-ordinated approach to problem solving. The nursing officer visited the States and Territories and discussed issues and problems with nurses in a wide range of hospitals, nursing homes, and community agencies. An assessment was made of the degree to which the problem solving mode was in use, and the extent to which systematic evaluation of aspects of nursing was implemented.

The nursing officer's survey revealed that in many larger hospitals and agencies there was interest in the problem solving approach, and initiatives had been taken towards implementation. In many instances, however, nurses needed a lot of help to understand the concepts and to develop the skills necessary for successful introduction of the approach. In 1979, the nursing officer recommended to RANF Federal Council a plan involving the convening of two national committees of nurses.

One committee, originally titled Standards and now styled Professional Development Committee (PDC), consists of three members selected by the nursing officer plus the nursing officer. This committee is responsible for preparing draft standards for discussion by members and subsequent adoption by RANF Federal Council. In addition the PDC is to prepare other documents as required to implement the national program and is to consult and advise the second committee.

The second committee, titled National Quality Assurance Committee (NQAC) is comprised of one representative from each RANF branch. The responsibility of the NQAC is to prepare and maintain a glossary of terms relevant to problem solving and quality assurance in nursing, and to plan and conduct workshops to introduce the concepts in each State/Territory. As the program progressed most RANF branches formed quality assurance committees to assist the branch NQAC member fulfil the responsibilities for conducting the workshops.

Several factors influenced the PDC when considering methods to be used in developing the standards. Committee members were anxious to achieve participation by as wide a range of nurses as possible. The method chosen should provide maximum opportunity for nurses to comment on, and suggest changes to, the work. The aim was to produce work that represented the views of, and was acceptable to, the majority of Australian nurses. By publishing the work in successive drafts these objectives could be achieved and widespread discussion and debate encouraged. Through discussion, it was thought, nurses would become more aware of the broader issues facing the nursing profession, rather than perceiving only the problems in the immediate work place. Broader understanding should promote unity and allow nurses to be more effective in dealing with issues.

FRAMEWORK OF THE RANF NATIONAL PROGRAM

From the outset it was obvious that an organising framework within which decisions could be made was essential. Since the aim was the establishment of a nursing care delivery system that makes it possible for nurses to assume professional responsibility for all aspects of nursing, the framework should allow nursing to be viewed from several perspectives.

The Structure, Process, and Outcome framework proposed by Donabedian (1969) was chosen to indicate the dimensions of the program. The model for quality assurance attributed by Lang (ANA, 1975) was adopted and used to define the steps in the quality assurance process. Use of this framework and model was considered to have several advantages. The framework of Structure, Process, and Outcome was relatively simple, well understood, and acceptable to a number of nurse educators and nurse administrators. In addition, it facilitated dealing with the work in logical steps.

Clearly defined steps seemed desirable for a number of reasons. RANF had limited financial resources to mount the program. The vast distances in Australia and the high cost of air fares precluded a massive education program. The representatives to the NQAC were achieving varying success in conducting the workshops. There was growing awareness that, although peer review was fundamental to problem solving and quality assurance, many nurses found the concepts both difficult to understand and very threatening.

The NQAC had defined peer review as 'the evaluation of performance of individuals or groups, by colleagues using established criteria' (RANF 1982a). This was seen to involve evaluation of the group by the group, as well as evaluation of an individual's performance by a colleague of more or less equal standing. It is believed that peer review in nursing may be conducted from three perspectives depending on the focus. Performance could be reviewed with the primary focus on the nurse, the patient, or the organisational characteristics of the nursing setting, and standards would be required for each.

Evaluation of structure: peer review with the focus on the organisational characteristics of the nursing setting

RANF is a member of the Australian Council on Hospital Standards (ACHS) which is concerned with accreditation of hospitals. RANF had contributed to the formulation of the nursing standards set out in that council's Accreditation Guide. In 1979 these standards were reviewed, redrafted, and submitted to ACHS for inclusion in the Accreditation Guide, 4th edition (ACHS, 1981). The revised standards were also published by RANF as *Standards for Nursing Divisions in Hospitals* (RANF, 1980).

During 1982 members were asked to review the Standards for Nursing Divisions in Hospitals and to suggest modifications and to determine if the standards could be applied in all nursing settings. Opinions received indicated that since the standards were based on recognised theories of organisation and administration they were applicable in any organised nursing setting.

The PDC subsequently prepared a guide, designed to help nurses achieve and maintain the standards to accompany the standards. The guide facilitates peer review and includes checklists for the major activities which need to be undertaken to ensure that the standards are achieved. A section designed to help in activities known to be causing most difficulty, such as development of philosophy and objectives, revision of job descriptions and specifications, was included. A selected bibliography of works considered helpful was also included.

It is not the intention that the guide should be slavishly followed. Rather, it is intended to help nurses make use of the resources available, and to look at the various ways organisation and administration theory may be applied in structuring a nursing division to promote effectiveness and efficiency in a particular setting.

The revised standards and the guide will be submitted to RANF Federal Council at the April 1983 meeting. If adopted the document will be published and be available by June 1983.

Evaluation of the process of nursing

Process was defined by the NQAC as 'the activities and interactions between those providing care and the recipient of care' (RANF, 1983). This definition is congruent with the notion that process may be reviewed from either of the two perspectives previously outlined. The primary focus may be on the practitioner or on the patient. These two aspects are considered to be complementary and equally important.

Review of process: peer review with the primary focus on the nurse

Review of nursing practice from this perspective has traditionally been termed performance appraisal.

Several factors influenced the decision to concentrate in the second phase of the program on evaluation of process with the primary focus on the nurse. The competencies of nurses registered under the various States/Territories Acts were diverse. There was a belief that citizens were entitled to expect the same level of competence regardless of the area in which they resided. Nationally agreed standards for practice were considered to provide a basis for the desired uniformity. In addition, such standards would indicate for society, students and nurses the competencies and practices which, in the opinion of nurses, make up nursing practice at the present time. Also influencing the decision to focus on the performance of the nurse was the knowledge that many nurses were dissatisfied with the current systems of performance appraisal and were eager to introduce criterion-referenced systems.

Formulation of standards for nursing practice

The PDC began the task of preparing standards for nursing practice in February 1981. In October 1981, the proposed draft standards were reviewed by a pilot group of 25 nurses. The group comprised educators, administrators, charge nurses and staff nurses. Modifications were made in the light of the many helpful suggestions made by the pilot group and the draft was published in 1982 (RANF, 1982a).

There was an excellent response from nurses who had individually studied the draft, but the majority of comments came from nurses who had worked in groups. The most frequent criticisms of the draft were that it would not be understood by nurses at 'grass roots level', and that the language was too jargonistic. Most correspondents supported the concepts contained in the draft but exhorted the committee to simplify the language. Several correspondents expressed the opinion that the document outlined competencies which could not be achieved by the majority of Australian nurses. Some asserted that what nurses actually did was dependent upon the orders of the medical practitioner and requested that the document be submitted to medical practitioners for an opinion. Again the standards were modified to take account of the comments and suggestions of the majority. Draft two, was published in October 1982 (RANF, 1982c).

In October 1982, a national conference to discuss standards for nursing practice was held. Some 400 nurses took part in the debate regarding the proposed standards. Conference participants passed resolutions calling upon the PDC to redraft the standards taking account of the conference discussions and to publish a third draft (RANF, 1983). Further comments were to be invited, and the PDC was directed to consider all comments received prior to March 1st, 1983. The resulting Standards for Nursing Practice are to be presented to RANF Federal Council at the April 1983 meeting.

A guide for implementing standards for nursing practice

Although the standards are considered applicable in every practice setting, it is recognised that the aspects of behaviour most suitable as indicators of compliance will vary from setting to setting and even within nursing units in a setting. Groups such as paediatric, orthopaedic, obstetric, domiciliary, community, rehabilitation, and critical care nurses will need to determine the most suitable indicators of compliance for their practice setting. The result should be standards for practice applicable to each specialty area.

The PDC is producing material designed to help nurses use the Standards for Nursing Practice to formulate such specialty standards. The guide is also intended to show nurses how the standards can be incorporated into specific job descriptions and used as a basis for performance appraisal.

Standards for Nursing Practice and the guide will be submitted to the RANF Federal Council at the April 1983 meeting. If adopted these documents will be published.

DEFINITION OF THE NATURE AND SCOPE OF PRACTICE OF THE NURSE

It is somewhat of a paradox that nurses are attempting to implement programs to assure quality in nursing practice while the nature and scope of nursing practice are still to be defined. Concurrently with the formulation of Standards for Nursing Practice, the PDC has been developing a document that defines the scope of nursing practice and the role and function of the nurse. This work is seen as a necessary adjunct to the Standards. Successive drafts of this statement have been published and modified according to members' comments (RANF, 1982b; 1982c). The resulting document will be presented to RANF Federal Council in April 1983.

Evaluation process: primary focus on the patient

The alternate focus of the process of nursing within the framework which has been adopted is the recipient of care, the patient. Within this focus, the process of nursing is inextricably tied up with the care received by the client. The criteria for review of process from this perspective is the care which is planned and delivered to the patient. These criteria should be found in the nursing care plan under the headings: nursing actions, nursing orders, nursing intervention, or whatever other title may be used to indicate the nursing care which has been considered appropriate to achieve the stated goals, objectives or expected outcomes of care. Thus, the nursing care plan becomes the primary source not only of the process criteria but also for the outcome criteria for review of care from the perspective of the patient. It is believed that this concept is valid regardless of whether the review is conducted on an individual basis as part of the ongoing process of care or whether the review is of the care of a representative sample of clients within the care setting.

Nursing standards for patient care: focus on the process and outcome of care

In order to review nursing practice with the primary focus on the patient on an agency-wide basis it will be necessary to develop 'nursing standards' for particular patient groups. Within any care setting there are groups of clients whose problems, actual or potential, are similar. The nature of the disability, surgical intervention, treatment regime, investigation or pathological process makes it possible to predict with a fair degree of accuracy the problems with which the patient will be faced and with which nursing help will be required.

A system can then be used to classify or categorise clients into

groups on the basis of common problems requiring nursing intervention. Nurses expert in the care of that particular group could itemise
the problems, agree on the outcomes that nursing care can achieve,
and detail the care most likely to promote that outcome. The result
would be care plans, prepared by nurse experts, which provide
guides to care and criteria for the evaluation of the process and
outcomes of care for particular patient groups. Within this system, it
is envisaged, for example, that critical care nurses would categorise
the problems commonly experienced by clients requiring intensive
care. The same process would be undertaken by nurses expert in
orthopaedic, domiciliary, paediatric, community, renal, medical,
surgical nursing and so on.

There is wide acceptance among nurses that the use of a problem
solving mode is fundamental to developing nursing standards for
patient care which include both process and outcome criteria. When
the concept of problem solving based upon a theoretical framework
for nursing is explained, most nurses perceive the usefulness of the
approach and are keen to use the concepts. Motivation is high.
However, in attempting to apply the concepts many difficulties
emerge.

It is from the educational background of most Australian nurses
that the fundamental difficulty arises. It is extremely difficult to
identify problems based on an assessment of a patient's requirements
for assistance to satisfy needs, if the system under which one learned
nursing was based on a medical model. This is the system under
which the majority of Australian nurses learned. Without resources
and a mentor, it is all too easy to become preoccupied with trivia and
tools and to lose sight of the original goals.

Even nurses experienced in using the problem solving approach
sometimes have difficulty applying the concepts learned in the educational settings in the practice setting. Most care plans developed
for learning are extremely long. Care plans in the practice setting are
not primarily designed for learning, they must provide a convenient,
workable system for continuity of care. Nurses find it difficult to
find time in a busy practice settings to write care plans. Consequently it is not unusual to find that, although the assessment has been
made, the care plan has not been developed beyond a checklist of
routine care.

Nurses who have been schooled in the medical model also find it
difficult to distinguish problems which require nursing intervention
from the medical diagnosis; they perceive the problem, for example,
as diabetes. In the absence of appropriate inservice education, ex-

emplars or role models in the practice setting, and help in writing care plans, nurses find the task overwhelmingly difficult and become frustrated. Contributing to this frustration is an incomplete understanding of the physiological and psycho-social basis for patients' problems. This lack of understanding results also from the use of the medical model in the education program and from the paucity of the physical and social science content of many of the courses.

These difficulties are understood by the PDC, but members believe that the approach being proposed for the development of nursing standards for patient care has advantages which outnumber the disadvantages. Each of the perceived advantages is discussed in the following sections.

Nursing standards as guides for care

The proposed nursing standards for patient care, established by experts in the clinical situation, will indicate what needs to be known and what needs to be done. The standards will provide nurses with the information where they need it — at the bedside. The standards are seen as guides which together with the nursing assessment will facilitate making clinical judgements. The nurse will be freed from the necessity of going through the long and time consuming process of handwriting the care plan for all the problems common and usual for that group. After completing the nursing assessment, nurses will be required to handwrite individual problems that have been identified, to state expected outcomes and to prescribe nursing care. This system will provide clear indicators of the practices for which the nurse is both responsible and accountable in the clinical situation and will overcome many of the difficulties discussed.

Nursing standards for patient care: relationship with patient classification systems

Australian nurses have been dissatisfied with current patient classification systems for determining the amount of care required by patients, and the category and mix of nurses which will prove effective from the perspectives of quality and cost. A frequent criticism is that current methods are more likely to take account of what is done rather than what should be done. Use of nursing standards for patient care as the indicators of the amount and type of care required may overcome this criticism.

Present patient classification systems fail to provide a consistent basis for determining the category of staff required to deliver the care planned for a patient. It is considered that the proposed nursing standards will allow the amount of care that requires the skill, knowledge and judgement of the registered nurse to be clearly identified. Conversely, the care that should not be entrusted to the second level nurse by virtue of training, agency, and statutory authority policies will be evident.

Cost of nursing care

Very little is known about the cost of nursing care for particular patients and patient groups. Nursing costs tend to be included in bed day costs. Nurses recognise their responsibility for cost effective care and the need for specific information that permits determination of actual costs and the most effective means to achieve stated outcomes. Until this information is available, nurses cannot assume full professional accounntability for the cost of nursing care. There will continue to be obstacles to providing quality, personalised care of known cost. Nursing standards for patient care, as described, should provide a basis for estimating costs and providing this needed information.

Nursing research

Australian nurses are acutely aware of the lack of data to link specifically certain nursing activities with outcomes in terms of a patient's health. The PDC members believe that the proposed nursing standards for patient care can be used as a base from which to hypothesise and test relationships between the process and outcomes of nursing care. Certainly, research will be needed to determine the reliability, validity, and sensitivity of the process and outcome criteria contained in the standards. Such research would also identify which criteria are indicators of quality and which are not. If these standards are used as a basis for research, much duplication of work could be avoided and the information so urgently required could be gathered on a co-operative basis.

Computers in nursing

The PDC members are aware of the rapidity with which computers are being introduced into patient care areas. Members are conscious that nurses, in many instances, are uncertain about the nursing data required. Nursing standards for patient care are readily adaptable to computerisation and are compatible with other nursing information such as nursing assessment and progress reports.

A working group is studying aspects of computerised patient data and nursing information systems with a view to making recommendations and developing guidelines which will help to ensure that, as far as possible, steps taken in developing standards and advocating quality assurance measures are compatible with future developments in the use of computers in nursing.

Categorisation of nursing diagnoses

An opinion is emerging among expert nurses that there is a need for a categorisation of nursing diagnoses. PDC members have been conscious and supportive of this view, but the urgency of the need to develop nursing standards for patient care has precluded this step. The approach being taken is believed to be compatible with the future development of a categorisation of nursing diagnoses.

A small number of Australian nurses do not agree that the three dimensional approach being advoacted by RANF has the advantages outlined. These nurses are anxious to implement nursing care reviews and are adopting already existing programs such as that advocated by Cabban et al (1980) or Haussmann et al (1976).

CONCLUSION

The RANF National Quality Assurance Program was commenced in July 1979. The committee members have worked consistently to achieve their objectives. The day to day progress has seemed slow but, taking an overall view, much has been accomplished.

As was anticipated, the progress and success of the NQAC members in conducting workshops has varied. Each NQAC representative is employed full time in an organised nursing service. It has been difficult sometimes for these nurses to obtain release from the agency to conduct the workshops. Often nurses have taken days from their annual leave entitlement to attend meetings or conduct the workshops. The greatest progress has been made where the NQAC representative has had support from the RANF branch office staff and from the Director of Nursing in the employing agency. In the State which has been most successful the workshops were organised in the main by a nursing officer employed by the RANF branch.

Throughout the program the emphasis has been on problem solving as a process, not on the tools that may be used. Emphasis has been placed on the cyclical nature of problem solving and on the use that can be made of the data obtained through systematic evaluation of nursing practice from the three perspectives. The interrelationship between identified problems and inservice education has been discussed. Efforts have been made to dispel the myth that most problems in nursing are due to lack of knowledge and to raise awareness of the other possible causes of unsatisfactory performance.

Despite the difficulties and frustrations in attempting to reconcile the many divergent opinions expressed, the RANF programs have been extremely useful. There is commitment to a problem solving approach. Many nurses believe it is necessary to continue to provide the opportunity for nurses to discuss the issues and explore the concepts rather than to adopt existing quality assurance programs developed by American nurses. The process of setting and agreeing on the standards is believed to be important in raising awareness about professional nursing and increasing unity in finding solutions to the many problems facing the profession. Already there is evidence of a professional growth among individual nurses and signs of a 'coming of age' of the nursing profession in Australia. This is perhaps the most important achievement.

A framework has been established to allow review of nursing practice from three perspectives. The effectiveness of the program will be demonstrated by the development and implementation of systematic reviews of nursing practice within this framework. A further indicator will be the extent to which data obtained from these reviews are used to improve nursing practice and achieve the major goal — an efficient and effective nursing care system in Australia.

REFERENCES

ACHS 1981 Accreditation guide for Australian hospitals and extended care facilities. Australian Council on Hospital Standards, Sydney

American Nurses' Association 1975 A plan for implementation of the standards for nursing practice. American Nurses' Association, Kansas City, Missouri

Cabban P T, Caswell J R, Adams A S 1980 Work study for hospitals. Community Systems Foundation (Australasia), New South Wales

Donabedian A 1969 Some issues in evaluating the quality of nursing care. American Journal of Public Health 59(10): 1833–1836

Donaghue S 1975 Goals in nursing education. Royal Australian Nursing Federation, Melbourne

Donaghue S 1977 Goals in nursing practice. Royal Australian Nursing Federation, Melbourne

Haussmann R K, Hegyvary S T, Newman J T 1976 Monitoring quality of nursing care, Part 2. U.S. Department of Health, Education and Welfare. Bethesda,

Maryland
RANF 1980 Standards for nursing divisions in hospitals. Royal Australian Nursing Federation, Melbourne
RANF 1982a Draft standards for nursing practice. Australian Nurses Journal 11(7): 15–17
RANF 1982b Draft proposal Nursing: a definitive statement. Australian Nurses Journal 11(11): 24–25
RANF 1982c Redrafted standards for nursing practice. Australian Nurses Journal 12(2): 18–19
RANF 1982d Nursing. Australian Nurses Journal 12(4): 14
RANF 1983 Standards for nursing practice: Draft 3. Australian Nurses Journal 12(6): 22–23

Choosing an appropriate method of quality assurance

Choosing an appropriate method of quality assurance in an institution first involves a critical evaluation of the existing methods. The aim of this chapter, therefore, is to briefly present the methods in current use in several countries and then to discuss the elements guiding the choice of a method appropriate to a particular situation. The methods in current use can be seen to evaluate the quality of care either partially or fully (globally).

METHODS CURRENTLY IN USE TO MEASURE QUALITY OF CARE

Partial evaluation of quality of care

In this case we can observe the following two trends:

1. creating one's own particular method; that is to say, adopting the nursing audit pattern;

2. applying an existing method such as that recommend by the American Nurses' Association and by the Registered Nurses' Association of British Columbia.

Nursing Audit Quality

The nursing department wishing to create its own auto-evaluation program may adopt the Nursing Audit Quality (Rubin et al, 1972; Gordon, 1980). This involves the following process:

— an Audit Committee composed of nurses (among whom those active in nursing management are essential) is set up and supported by the administration;
— this Committee identifies a topic known to present nursing care problems;
— the Committee sets goals regarding this audit topic developing and ratifying measurable criteria in a check-list with the common

assent of the staff nurses involved in the audit. The standard (0–100%) must be specified;
— to measure the practice against the standard the check-list, used as a work sheet, is filled in on the spot by a member of the Committee;
— the data are analyzed and the Committee determines the type of deficiency (knowledge, documentation, practice);
— the results are presented to the nursing staff and, if the latter is responsible for errors, recommendations for corrective action can be made. The time frame for completion of corrective action should be specified;
— a new audit on the same topic is then arranged to evaluate the corrective actions;
— the audit results are then reported to the administration and to the hospital governing body (Benedikter, 1977; Doughty & Mash, 1977; Breuil, 1981; de Villermay, 1982).

It is possible to apply this method of evaluation in a relatively short time because the number of criteria used for each topic is small (20 to 30). The results are quite easy to compute and the Audit Committee may quickly take corrective action. The topics are very often process-oriented and based on a prospective approach. The cost involved is also quite reasonable; it does not require an increase in either staff or equipment.

While this method is self-stimulating for nursing staff and, indirectly, permits care standardization it does not call into question the reliability of the investigation nor the sensitivity and validity of the criteria (Fey & Gogue, 1980; Jacquerye, 1983). The Nursing Audit Quality method is in favor in Europe, especially in France.

Method of the American Nurses' Association (ANA) and of the Registered Nurses' Association of British Columbia (RNABC)

The ANA since 1973 has, in different publications, developed recommendations to help fulfil the profession's obligation to provide clients with quality nursing services (ANA, 1976a; 1976b; RNABC, 1977; Gordon, 1980). These recommendations present standards of nursing practice and a Model for Quality Assurance in Nursing as guides in the development of an evaluation system in any setting. The components of the Model, presented in an open and circular diagram, are as follows:

1. Identify values — this means that standards and criteria reflecting societal and professional values must be established;

2. Identify structure, process and outcome standards and criteria. Review committees should be formed within the committee structure of the institution and should apply administrative sanctions. These committees must decide which types of criteria they will use, depending on the purpose and target of the program;

3. Gather measurements needed to determine the degree of attainment of standards and criteria by direct observation of patients, interview, questionnaire, performance observation or prospective or retrospective review of records;

4. Make interpretations about the strong and weak parts of the program;

5. Identify possible courses of action: continuing education, in-service education, peer pressure, administrative changes, self-initiated changes, etc.;

6. Choose a course of action;

7. Take action;

8. Re-evaluation — At this point the cycle begins again. Each time actions are taken the progress of nursing practice needs to be remeasured and reassessed.

The ANA also proposed sets of criteria for specific populations of patients previously used in a pilot test for validity and suitability. In addition to the basic standards the ANA has published standards for various fields of nursing practice such as: psychiatry, mental health, gerontology, orthopedics, cardiovascular disease. To comply with the federal requirements of the Professional Standards Review Organization (PSRO) the ANA specially recommended use of outcome criteria and the retrospective approach (United States Department of Health, Education and Welfare, 1977; Egelston, 1980; Inzer & Aspinall, 1981). Because of the influence of other standard setting agencies it is difficult to separate out the extent of the nursing profession's responsibility for result. But using this approach makes it difficult to identify outcomes attributable to nursing. Jacquerye (1983) points this out in respect to the ANA set of standards — Management of the surgical patient: cholecystectomy.

In conclusion, this method responds to two specific needs. It provides a simple framework useful to practitioners and is in keeping with professional and legal requirements.

Global evaluation of quality of care

As far as we know there are, at present, five methods of evaluating quality of care globally:

1. Nursing Audit of Phaneuf (1972, 1976);
2. The Quality Patient Care Scale (QUALPACS) of Wandelt & Ager (1974);
3. The Rush-Medicus (Jelinek et al, 1974);
4. The development of criterion measures of nursing care quality (Horn & Swain, 1977);
5. Méthode d'Appréciation de la Qualité des Soins Infirmiers (MAQSI) published by the Order of Nurses of Quebec (Chagnon et al, 1982a, 1982b).

Nursing audit

Phaneuf (1972, 1976) created a method to evaluate the quality of care on a monthly basis in any institution. The method is essentially based on retrospective audits of a sample of 50 closed records of patients who were discharged from the hospital during the previous month. The examination of each record asks the reviewer to check 50 items. These are subdivided in relation to seven functions of nursing as delineated under the legal functions of nursing. The seven functional areas used as a framework are:

— application and execution of the physician's legal orders;
— observation of signs and symptoms and reactions;
— supervision of the patient;
— supervision of those participating in care (except the physician);
— reporting and recording;
— application and execution of nursing procedures and techniques;
— promotion of physical and emotional health by direction and teaching.

The examiner reading the records has to check the box of choice (yes, no, uncertain, not applicable) for every one of the 50 items. The total for each of the function categories and the total score for the seven functions are then calculated. The results can then be placed in five possible numerical ranges corresponding to the nominal ranges: excellent, good, incomplete, poor, unsafe. The results are then entered on to a sheet to construct a profile of excellence. Each of the seven functions of the actual score is compared to the ideal score and the difference is an index of performance. By examining the graph the nursing administrator can identify the needs, make recommendations, implement a plan of action and prepare for a new evaluation.

Phaneuf proposed the setting up of an Evaluation Committee which would consist of at least five nurses well known for their clinical competence. Each member would systematically evaluate ten records monthly. A common objection is that the quality of care is measured only on the basis of a written record, but now a great deal of nursing care never gets recorded at all so that auditing can never be complete (Wandelt, 1973; Jacquerye, 1983). However, in spite of this, the Nursing Audit of Phaneuf is an easy and economical method. It gives a global retrospective picture of the quality of nursing care.

The quality patient care scale (QUALPACS)

The QUALPACS is a 68 item scale prepared for measuring the quality of care received by patients either from direct nurse-patient interactions or from interventions on behalf of the patient (Wandelt & Ager, 1974). The scale is outlined in six areas as follows:

a. care received directed toward meeting psychosocial needs of the patient as an individual

b. care received reflecting recognition of the patient's psychosocial needs as a member of a group

c. care received meeting physical needs

d. care received meeting both psychosocial and physical needs at once

e. dealing with communication

f. dealing with professional responsibility

The QUALPACS design is appropriate to measure care. It provides for a controlled random selection of five or six patients to represent the level of quality of care being received by patients in the unit sampled. The use of the scale implies direct observation of a patient as he is receiving care and the rating of items on the scale. It results in a mean score for this particular patient, and in turn a mean score for that unit.

The possible ranges of score correspond to best care (5), between (4), average care (3), between (2), poorest care (1). The expected norm is 'best care' for one or all the patients. This method is well presented in guide form ready to be used on the spot. However as, in our opinion, only one-third of the criteria are specific to the nursing profession, we think it would be difficult to apply corrective actions (Jacquerye, 1983).

Rush-Medicus

The authors of the method *Monitoring Quality of Nursing Care* (Haussmann et al, 1976) are nurses from the University of Rush-Presbyterian-St Luke's Medical Center, the Rush College of Nursing and statisticians from a private business firm, the Medicus Systems Corporation. The framework of the method is based on the nursing process model. The nursing process is the comprehensive set of nursing activities performed in the delivery of a patient's care (Jelinek et al, 1974; Hegyvary & Haussmann, 1975; Haussmann & Hegyvary, 1976; Haussmann et al, 1976).

Each of the nursing process areas was operationally defined to finally give the following plan:

Revised objectives and subobjectives structure
1.0. The plan of nursing care is formulated
1.1. The condition of the patient is assessed on admission.
1.2. Data relevant to hospital care are ascertained on admission.
1.3. The current condition of the patient is assessed.
1.4. The written plan of nursing care is formulated.
1.5. The plan of nursing care is co-ordinated with the medical plan of care.
2.0. The physical needs of the patient are attended
2.1. The patient is protected from accident and injury.
2.2. The need for physical comfort and rest is attended.
2.3. The need for physical hygiene is attended.
2.4. The need for a supply of oxygen is attended.
2.5. The need for activity is attended.
2.6. The need for nutrition and fluid balance is attended.
2.7. The need for elimination is attended.
2.8. The need for skin care is attended.
2.9. The patient is protected from infection.
3.0. The nonphysical (psychological, emotional, mental, and social) needs of the patient are attended
3.1. The patient is oriented to hospital facilities on admission.
3.2. The patient is extended social courtesy by the nursing staff.
3.3. The patient's privacy and civil rights are honored.

3.4. The need for psychological-emotional well-being is attended.

3.5. The patient is taught measures of health maintenance and illness prevention.

3.6. The patient's family is included in the nursing care process.

4.0. Achievement of nursing care objectives is evaluated

4.1. Records document the care provided for the patient.

4.2. The patient's response to therapy is evaluated.

5.0. Unit procedures are followed for the protection of all patients

5.1. Isolation and decontamination procedures are followed.

5.2. The unit is prepared for emergency situations.

6.0. The delivery of nursing care is facilitated by administrative and managerial services

6.1. Nursing reporting follows prescribed standards.

6.2. Nursing management is provided.

6.3. Clerical services are provided.

6.4. Environmental and support services are provided.

(Haussmann et al, 1976, p 7)

In the plan, approximately 250 criteria are distributed in two parts, each grouped in six classes called objectives; these six objectives are distributed in 28 categories called subobjectives. The subobjectives are thus receiving a series of specific criteria related to the same subject.

All the criteria cannot be systematically applied to all patients. For this reason, a workload system classifies the patients as type 1, 2, 3 or 4, meaning self, partial, complete and intensive care respectively. In addition, the criteria are applicable to medical, surgical, pediatric nursing units and newborn nurseries and recovery rooms. Each criterion is coded according to the type of patient to which it most likely would apply and to the type of area. Approximately 170 criteria are then patient specific; the remaining 50 are unit specific. Not all the criteria are used for evaluating the nursing process with regard to each patient and unit setting. Rather, the master criteria list is placed in a computer file from which subsets of criteria (30 to 50, depending on patient type) are systematically selected and grouped into worksheets. The selection process is such that each worksheet covers the entire spectrum of criteria but selects randomly some of them within each subobjective.

Quality in any nursing unit is evaluated on the basis of a review of

10% of one month's patient census (12 to 20 patients, depending on unit occupancy and length of stay). A co-ordinator is appointed to define a master schedule of the various observations to be made by shift in each unit by the nurse interviewers. Specific patients are randomly selected from the unit just prior to the actual observations. Once patients have been identified, their type and classification are ascertained and appropriate worksheets are selected for use. This method is thus only prospective and based on observations, interviews, and examination of current records. At the end of the month, a computer program produces quality indexes for the 28 subobjectives.

Each subobjective index is the average of the criterion scores within the subobjective. Each criterion score is the rate of positive responses to the maximum possible positive responses, based on the number of valid observations for the criterion that month. Indices for objectives are computed as average values of the subobjective scores within a given objective. The data also can be interpreted by types of shifts, units and patients. Programs exist for generating worksheets from the master list, for editing data and for developing quality indices. Any hospital with a data processing system or access to one can set up these programs and operate the monitoring system (ANS COBOL).

This method was tested in two pilot hospitals and a careful examination of the reliability, the validity and sensitivity of the criteria was made. This method provides a global picture of the quality of care incorporating all useful elements of existing methodologies. However, in our opinion, this method is quite difficult to apply because it implies a precise fulfillment of the workload system and a prospective collection of data based on interviews and observations of the family and the staff and the survey of the records (Jacquerye, 1983).

Development of criterion measures of nursing care quality

Horn & Swain (1976; 1977) described a project which demonstrated a method for evaluating the outcomes of nursing care process as reflected by the patients' health status (Howe, 1980). This work was only partially tested. Because of its extensive and complex conception and development we have to limit the presentation to the essentials.

The authors have adopted as a framework Orem's conception of care. Nine categories of health status dimensions are developed from

Orem's description of self-care requirements related to air; water/ fluid intake; food; elimination of body wastes; rest/activity/sleep; solitude, social interaction and productive work; protection from hazards; normality; and health deviation. These nine aspects of health provide the necessary framework for organizing nursing care outcome. For further specificity in defining the domain of nursing care, each of the categories of health status is subdivided into four areas:

I. Evidence that the requirement is met;

II. Evidence that the person has the necessary knowledge to meet the requirement;

III. Evidence that the person has the necessary skills and performance abilities to meet the requirement;

IV. Evidence that the person has the necessary motivation to meet the requirement.

This nine-by-four framework for describing the scope of nursing's responsibility is a device for the generation of patient outcome variables. The nurse staff selects seven of the nine categories for initial forms (air, water) and proceeds to list all the variables indicated by the domains I-IV for each category. Variables are thus generated that demonstrate the degree to which the patient's universal demand for air has been met (domain I), the adequacy of knowledge necessary to meet his/her need for air following hospitalization (domain II), and his/her ability to and motivation to manage effectively self-care related to getting air (domain III and IV).

Normal or optimum levels of achievement are also described for each variable generated. In the health deviation category, ten subsets of dimensions associated with health deviation and related therapies are generated (for examples, i.v. and wound observations, medication knowledge and injection performance).

The measuring instrument is composed of 539 validated items of which 414 were pretested for reliability. All the 539 cannot be applied to each patient. The authors classify the patients into 90 groups according to their medical diagnosis and criteria are specific to these groups.

Data collection is designed to use the patient as the primary source of data, so direct physical observations and interviews are the most common types of measurement techniques developed. Approximately 1 hour is devoted per patient. The measures are designed so that staff nurses could readily be trained to do the observations and interviews and so that the equipment required would be available in most hospitals or could be obtained without great expense. The implementation of the method is preferably in the hands of a co-ordinator. This method, to our knowledge, has never been applied in entirety so it is difficult to have a real idea of the global picture of the expected results. As in all the outcome orientations, we find it difficult to evaluate the specific part that belongs to nursing care.

The interest of this method lays especially in the wealth of criteria we can be inspired to create through using a partial method based on specific patients or specific nursing actions (Jacquerye, 1983).

*Méthode d' appréciation de la qualité des soins infirmiers (MAQSI)**

At the end of 1982 the Order of Nurses of Quebec published a quality assurance method (Chagnon et al 1982a; 1982b).

This method, based on seven measurements, is applicable to six groups of hospitalized patients chosen according to an adaptation of Kessner's selection criteria. The norms and criteria are inspired by Orem's conceptual model and are outcome oriented. The report details the terms and conditions necessary to apply this method (prospective approach, education program for the interviewers, system of data coding and analyzing).

This method was tested in six hospitals for the validity of content, reliability and significance, and then tried out on a experimental basis in two hospitals. The method has the advantage of being applicable to populations commonly observed in all hospitals: major abdominal surgery cases 48 hours after the operation, and at time of discharge; the newborn; the primigravida; the diabetic patient; the hospitalized child with pulmonary problems and the colostomy patient. This method, the most recent one, is quite clear and relevant. It can be used readily and does not offer any particular difficulty and in such a system the part played by nursing seems to be better defined. An interesting feature also is that the method would be easy to use in partial evaluation of care.

ELEMENTS GUIDING THE CHOICE OF AN APPROPRIATE METHOD

By examining all these methods, it is clear that considering the

---

* I particularly would like to thank Monique Chagnon for her advice for the MAQSI's method at the International Symposium on the Evaluation of Nursing Care held in Brussels, Cliniques Universitaires de Bruxelles, Hopital Erasme, February 17–18, 1983.

inherent advantages and disadvantages of each method the choice has to take into consideration the particular features of each institution. Since the elements guiding the choice are numerous we propose to develop principally the following:

1. the objectives of the management and the size of the institution;
2. the time required for starting;
3. the possibilities of investment;
4. the time required for corrective actions.

Objectives of the management and size of the institution

A program evaluating the quality of nursing care can respond to two clear and often complementary finalities: first, to insure that within the institution itself there is attainment of a quality that is labelled in the nursing process and/or, second, to comply with an external control either legal or issuing from the private insurance companies (Diddie, 1976; Felton et al, 1976; Lang, 1976; Selvaggi et al, 1976). Whatever the size of the institution a method of partial evaluation can be chosen. Partial methods have the advantage of being adaptable to local objectives and needs. The ANA's method corresponds particularly to the recent requirements of the Joint Commission on Accreditation of Hospitals (JCAH), which recommends that institutions should work out specific problems of quality of care (ANA 1976a; 1976b; Sward, 1976; Egelston, 1980; JCAH, 1982).

The Rush-Medicus and Horn and Swain methods are above all applicable to large institutions (at least 500 beds) or to medical centers. They also take into account external controls especially recommended in the years 1975–1979 (Lang, 1976).

The QUALPACS, the Nursing Audit of Phaneuf, the Quality Nursing Audit and the MAQSI's method seem to respond to an internal professional control whatever the size of the institution.

Time required for starting

If an institution wants to introduce within a short time a program of quality of nursing then the two methods of partial evaluation are useful. They are, however, facilitated by two more conditions: the general use of the nursing record and the routine using of care standards (Jacquerye, 1983). The QUALPACS requires more time to be applied. The Horn and Swain and the Rush-Medicus Methods require before application a quite heavy structure (workload classification, need of co-ordinator ...), a serious adaptation period and, particularly for the Rush-Medicus, the daily use of the nursing record. Finally the MAQSI method is the best ready-to-use in any hospital.

Possibilities of investments

Investment in time and personnel

The lowest costing methods are, without any doubt, the two retrospective methods ANA and the Nursing Audit of Phaneuf and, as third choice, the prospective Quality Nursing Audit. The four others (MAQSI, QUALPACS, Rush-Medicus and Horn and Swain) make greater demands in respect to personnel, particularly the co-ordinator and the interviewers. We have to note that these additional responsibilities do not necessarily require more personnel but they will keep some nurses away from usual routine and usual work.

Investment in equipment and material

It seems that the Rush-Medicus method demands most investment; it requires the purchase of a computer program, thus implying routine use of a computer in the institution.

In the QUALPACS and the Nursing Audit of Phaneuf the data analysis would be facilitated by the use of a computer, especially if the size of the sample is large.

The MASQSI presents in the users' manual the whole coding system for computers (80 columns) so that it is the most attractive. It is more difficult to give an opinion on the global method of Horn and Swain because it was not fully applied.

Time required for corrective actions

The methods of partial evaluation as well as the QUALPACS and the MAQSI allow corrective actions with a relatively short delay. On the other hand, the results of the evaluations from the Nursing Audit of Phaneuf and the Rush-Medicus are obtained monthly. This fact extends the delay of corrective actions; nevertheless, it is possible to take out parts of these global methods and then, according to the needs, apply corrective actions immediately.

CONCLUSION

The most important purpose of an evaluation method is to improve the quality of care. To us it seems relevant to apply first the following steps:

1. to select a method of partial or a part of a global evaluation;
2. to apply this method in a pilot unit;
3. to examine the potential improvements.

If the balance is quite positive, you could establish this method as a standard and extend its application to other units. Otherwise, you could choose another method of partial evaluation or other groups or criteria of a global method following the same steps. It is only after a real experience of all the difficulties of an application of a partial evaluation that an institution is ready to apply a global method. When the latter is employed it also should be analyzed concerning the generated improvements.

REFERENCES

American Nurses' Association (ed) 1976a Quality assurance workbook. ANA, Kansas City

American Nurses' Association (ed) 1976b Guidelines for review of nursing care at the local level. ANA, Kansas City

Benedikter H 1977 From nursing audit to multidisciplinary audit. National League for Nursing, New York

Breuil M T 1981 Le nursing audit ou l'évaluation des soins infirmiers à l'Hôpital américain de Paris. Soins 26(6): 40–48

Chagnon M, Lange-Sondack P, Arlot-Tovel D 1982a Méthode d'appréciation de la qualité des soins infirmiers Tome 1 rapport de recherche. Ordre des Infirmières et Infirmiers du Québec, Montréal

Chagnon M, Lange-Sondack P, Arlot-Tovel D 1982b Methode d'appréciation de la qualite des soins infirmiers Tome 2 Manuel de l'usager. Ordre des Infirmières et Infirmiers du Québec, Montréal

de Villermay D 1982 Pour améliorer la qualité des soins: l'audit. Revue de l'Infirmière 32(1): 65–69

Diddie P J 1976 Quality assurance — a general hospital meets the challenge. Journal of Nursing Administration 6(4): 6–8+

Doughty D B, Mash N J 1977 Nursing audit. Davis, Philadelphia

Egelston E M 1980 New JCAH standard on quality assurance. Nursing Research 29(2): 113–114

Felton G, Frevert E, Galligan K, Neill M K, Williams L 1976 Pathway to accountability: implementation of a quality assurance program. Journal of Nursing Administration 6(1): 20–23

Fey R, Gogue J M 1980 La maîtrise de la qualité. Les éditions d'organisation, Paris

Gordon M 1980 Determining study topics. Nursing Research 29(2): 83–87

Haussmann R K, Hegyvary S T 1976 Monitoring nursing care quality. Journal of Nursing Administration 6(9): 3–9

Haussmann R K, Hegyvary S T, Newman J F 1976 Monitoring quality of Nursing care Part II — Assessment and study of correlates. Department of Health, Education and Welfare, Bethesda, Maryland

Hegyvary S T, Haussmann R K D 1975 Monitoring nursing care quality. Journal of Nursing Administration 5(5): 17–26

Horn B J, Swain M A 1976 An approach to development of criterion measures for quality health care. In: American Nurses' Association (ed) Issues in evaluation research. ANA, Kansas City, p 74–82

Horn B J, Swain M A 1977 Development of criterion measures of nursing care. Department of Hospital Administration, School of Public Health, the University of Michigan Ann Arbor (Unpublished)

Howe M J 1980 Developing instruments for measurement of criteria. Nursing Research 29(2): 100–103

Inzer F, Aspinall M J 1981 Evaluating patient outcomes. Nursing Outlook 29(3): 178–181

Jacquerye A 1983 Guide de l'evaluation de la qualité des soins infirmiers. Centurion, Paris

Jelinek R C, Haussmann R K D, Hegyvary S T, Newman J F 1974 A methodology for monitoring quality of nursing care. Department of Health, Education and Welfare, Bethesda, Maryland

Joint Commission on Accreditation of Hospitals (ed) 1982 Accreditation manual for hospitals. JCAH, Chicago

Lang N M 1976 Issues in quality assurance in nursing. In: American Nurses' Association (ed) Issues in evaluation research. ANA, Kansas City, p 45–46

Phaneuf M C 1972 The nursing audit — Profile for excellence. Appleton-Century-Crofts, New York

Phaneuf M C 1976 The nursing audit – Self-regulation in nursing practice, 2nd edn. Appleton-Century-Crofts, New York

Registered Nurses' Association of British Columbia (ed) 1977 Quality assurance manual. Vancouver British Clumbia

Rubin C F, Rinaldi L A, Dietz R R 1972 Nursing audit — nurses evaluating nursing. American Journal of Nursing 72: 916–921

Selvaggi L M, Eriksen L, Kean P, MacKinnan H A 1976 Implementing a quality assurance program in nursing. Journal of Nursing Administration 6(7): 37–43

Sward K M 1976 Some perspectives on ANA and quality assurance. In: American Nurses' Association (ed) Issues in evaluating research. ANA, Kansas City, p 29–44

Wandelt M A 1973 Systems and tools for evaluation of nursing care/past and present. In: American Nurses' Association and American Hospital Association (eds) Quality assurance for nursing. ANA and AHA, Kansas City, p 18–26

Wandelt M A, Ager J W 1974 Quality patient care scale. Appleton-Century-Crofts, New York

U.S. Department of Health, Education and Welfare (ed) 1977 PSRO — Technical assistance — document 9. USDHEW Health Care Financing Administration Health Standards and Quality Bureau

Research to Change Outcomes of Care

Staffing methods — implications for quality

The notion that nursing resources are crucial to the quality of patient care seems commonplace and unexceptional. Yet widespread acknowledgement of the importance and use of reliable, valid and sensitive measures for determining nursing resources and quality of care has until quite recently been both uncommon and exceptional. While other chapters in this book are focused on issues relating to the measurement of the quality of care, this chapter is focused specifically on the methods employed for the determination and allocation of nursing resources.

For the purpose of analysis here, nursing resource determination is viewed from the perspective of staffing at the unit or ward level. Firstly, in the methods presented here it is assumed that identified requirements at this level may be aggregated to provide resource requirements at the institutional level and ultimately at the regional and/or national level; but more importantly variations *between* units are considered too significant to be ignored, a danger that pervades contrasting methods that establish global standards for the purpose of extrapolation to the unit level.

A perspective from which to view the scope of nurse staffing is provided by Aydelotte (1973a). She suggests that a comprehensive nurse staffing program contains four elements:

1. A precise statement of the purpose of the institution and the services a patient can expect from it, including the standard and characteristics of the care;
2. The application of a specific method to determine the number and kinds of staff required to provide the care;
3. The development of assignment patterns for staff from the application of personnel guidelines, policy statements, and procedures;
4. An evaluation of the product provided and judgement reflecting the impact of the staff upon quality (p. 60).

The major focus in this chapter will be on the second and fourth elements: the methods employed for determining the numbers and kinds of staff required to provide care; and evaluation, the link between staffing resources and the quality of care. The chapter begins with the description of a conceptual framework for nurse staffing. The historical development of staffing methods is discussed followed by a descriiption of staffing approaches that are based on patient classification. Several approaches to measuring nursing care time are identified and the application and monitoring of staffing methods is discussed. The chapter concludes with comments relating to the link between nurse staffing and the quality of patient care.

CONCEPTUAL FRAMEWORK FOR NURSE STAFFING

The elements of a staffing program set forth by Aydelotte form a useful basis for the development of a conceptual framework for nurse staffing. Several investigators have attempted to illustrate graphically the relationships among variables affecting staffing. While there are differences amongst various frameworks there appears to be agreement on the theme that nurse staffing occurs within a dynamic system, influenced by a great number of interrelated factors over and above patients' requirements for nursing care (Nadler & Sahney, 1969; Giovannetti, 1973; Jelinek & Dennis, 1976; U.S. Department of Health, Education and Welfare, 1978; Young et al, 1981). Figure 6.1 represents one perspective for viewing nurse staffing at the nursing unit or ward level (U.S. Department of Health, Education and Welfare, 1978). The framework is premised on the notion that there is a continuous negotiation between staff and patients to reach agreement on the services to be provided. It should be noted that the definition of staffing reflected in the framework was as follows:

> The provision of the appropriate amount and type of care by persons possessing the requisite skills to the largest number of patients possible in the most cost-efficient and humanly effective manner consistent with desired patient outcomes and personnel needs for satisfaction (p. 67).

Figure 6.1 displays a wide range of factors potentially affecting a staffing program at the ward or unit level, the relationship between and the direction(s) of effects of the factors. The right side of Figure 6.1 specifies the various events that may arise to cause a discrepancy between the staffing plan and the care received by patients. Because

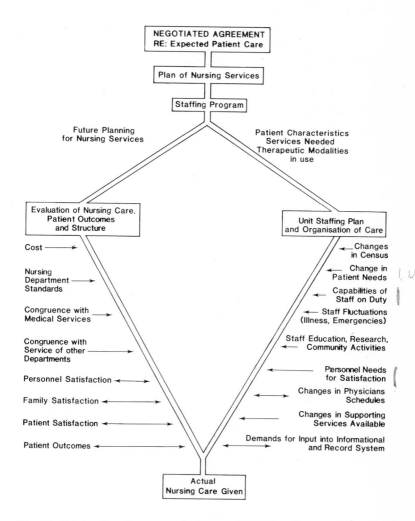

Fig. 6.1   Relationship of factors entering into provisions of nursing care at the unit level

little can be done to alter the occurrence of most of these events, such as changes in patients' needs and emergencies, their influence is depicted as operating in one direction. Personnel needs for satisfaction and demands for input into informational and record systems, however, are subject to alteration and thus are depicted as having a two-way, or interactive, relationship with the system.

The left side of Figure 6.1 highlights the interdependency of other groups within the hospital in the process of evaluation of the actual nursing care given. As indicated, the evaluation of care involves multiple structure and outcome variables. The open system features of this framework allow for the evaluation of the nursing care to feed back into planning for future nursing services, the staffing program, the overall plan of nursing services, and the 'negotiated agreement'.

For the purposes of this chapter, discussion will be limited to only one of the factors that affects staffing, namely patients' requirements for nursing care. It is understood however that nurse staffing is a complex issue and consideration and analysis of only one aspect, however significant, cannot be expected to produce a definitive resolution to all staffing questions.

Before proceeding with the discussion of staffing methods, it is important to distinguish between 'staffing levels' and 'staffing patterns'. Young et al (1981) in their review and critique of the literature relating to factors affecting nurse staffing, provide the most comprehensive definitions. Staffing levels are defined as:

> ... the gross number of nursing personnel designated for a given area, such as the inpatient nursing unit or a medical or surgical service. In many instances, the term may be operationalized to refer to the total nursing hours required per shift as indicated by one of the many staffing methodologies based on patient classification and workload estimation. Staffing levels may also be presented as the number of full-time equivalents. Usually, no attempt is made to specify skill levels (p. 12).

In contrast, staffing patterns are described as:

> ... the mix or ratio of professional to nonprofessional nursing personnel for a specific nursing unit, nursing service, or facility. To a large extent, this term is a refinement of the specification of staffing levels in that a more detailed indication of skill levels is provided; in many studies, staffing patterns refer to the configuration of nursing personnel, such as the number of registered nurses, licensed practical nurses, and aides required

to respond to the care needs of a given patient population (p. 12).

STAFFING METHODS

Historical developments

Historically, the determination of nurse staffing has been based on arbitrary methods culminating in global standards which in turn have been greatly influenced by the dual forces of pressure and precedents. Budget constraints and market conditions have served to exert the pressures while the precedents stem from historical practices. The effects of such practices on the provision of nursing care have never been well documented, no doubt because of the difficulty inherent in measuring the quality of care. Nevertheless, the professional judgement of many nurses in reference to staffing practices has culminated in statements reflecting either 'feast or famine' and concerns regarding their ability to provide safe nursing care.

Such comments, however, are frequently ignored on the grounds that they are emotionally charged and void of empirically derived evidence. The situation is somewhat ironic. The practice of nursing depends greatly upon the professional judgement of qualified nurses. Yet, that same judgement when applied to the resources required to carry out nursing practice is rarely considered as anything more than subjective and self-serving. Professional judgement when focused on the components of care provided to the nation never achieves the level of debate that occurs when this same judgement is linked directly to the pocketbook of the nation. It would appear that as long as professional nurses are employed by institutions to provide nursing care rather than serve as independent practitioners, nursing practice will be seriously constrained by the economic considerations that prevail.

The practice of using global standards on which to base staffing levels remained, until the early 1960s, virtually unchallenged. In some instances the standards were regional, in others they were altered according to hospital size and/or specialty area. Whatever the standards, there is little evidence to suggest that they were based on either formal quantitative or qualitative study or that they were operationally linked to explicit standards of nursing practice.

Ironically, one of the earliest attempts to study nursing care quantitatively served to perpetuate the use of global standards. A survey of 50 selected hospitals in New York City revealed that the median number of hours of bedside nursing care was between 3.4 and 3.5 (National League of Nursing Education, 1937). As a result of the study, the figures of 3.4 to 3.5 total nursing hours per patient day were recommended '... not because they are known to be right but because ... it would appear to be a practical recommendation for the present.' Along with this recommendation, the investigators identified the need for further information based upon sound investigation of the factors essential for organizing and evaluating hospital nursing service, and for determining the right number of hours of nursing for the various types of ward patients. Little attention was paid to the suggestions for further study and 3.4 to 3.5 hours per patient day became widely accepted across North America as a staffing standard. Almost 30 years later, a survey of randomly selected hospitals in Canada revealed that the standard most commonly accepted for estimating nursing staff requirements was 3.5 hours of care per patient day (Canadian Nurses Association, 1966), as was the case for many years in the United States (Abdellah & Levine, 1979).

The history of staffing is both fascinating and revealing, and many excellent accounts of the early staffing studies can be found in the literature (Aydelotte, 1973a; Baar et al, 1973; Jelinek & Dennis, 1976; Abdellah & Levine, 1979; Buchan, 1979; Giovannetti, 1979; Young et al, 1981). Of particular interest is the fact that many of the issues of today were highlighted in the early studies. For example, two studies by the National League of Nursing (1948, 1949) discussed the concept of grouping patients into clusters and reported that the patients' degree of illness did not necessarily reflect the intensities of nursing care needed. Wright (1954) reported that patients in an acute or critical state should be almost exclusively under the care of professional nurses. George & Kuehn (1955) noted that a different kind of care was often needed for patients classified as mildly ill since it was during the convalescent phase that the major part of health teaching was undertaken, and could be of such magnitude as to represent more time than that required for treatments and medications ordered for the acutely ill patients. They went on to report that patients with extensive emotional reactions, even though they may be medically classified as moderately ill, could require as much or even more nursing time than those whose general condition was considered to be more serious.

While the major focus of most of the early staffing studies was on staffing patterns rather than staffing levels, they signify the beginning of the separation between the medical model stressing the seriousness of the diagnoses and the nursing model stressing the magnitude of nursing care time. Some of the early studies did establish that the hours of nursing care were different for different patients (Bernstein, 1953; Wright, 1954; George & Kuehn, 1955). One may speculate, however, that had the distinction between the medical model and nursing model been more clearly defined, global standards such as average hours of care per patient per day would have been challenged at a much earlier date.

The concrete basis for the challenge to global staffing standards came primarily from the work at the Johns Hopkins Hospital. Connor (1960) and Wolfe & Young (1965a, 1965b) were the first to show quantitatively what had been known experientially and intuitively for years: some patients require more care than others. In addition, several other significant findings were revealed. The first, was the relationship between nursing workload and patient census. The investigators demonstrated that nursing workload was not a function of gross census alone, but rather of the number of patients in each category of care present on the ward. Second, a wide variation in total nursing workload existed from day to day and shift to shift. Third, the variation in nursing workload was independent from ward to ward. Finally, the main determinant of total nursing workload was the number of Class III or intensive nursing care patients.

In the past two decades, similar findings have been replicated in hundreds of studies throughout North America and Great Britain. The result has been a proliferation of classification schemes and staffing methods. In some instances the classification instrument developed by Connor was modified by hospital employed nurses interested in implementing more rational staffing policies; in others, by proprietary agencies. In virtually all instances, the work at the Johns Hopkins Hospital served as the template for further developments.

The number of facilities using patient classification has risen substantially during the last decade. In 1978, it was estimated that approximately 1000 hospitals in the USA were using a patient classification instrument for nurse staffing (Giovannetti, 1978). In light of the endorsement and encouragement of the Joint Commission for Hospitals Association in the USA, it is assumed that almost all hospitals with a bed capacity greater than 200 are at some stage of implementing patient classification. In Canada, the results of a survey conducted by the Canadian Hospital Association in 1982, and based on a response rate of 62%, reported that approximately 25% of the hospitals in Canada were using a patient classification system (Canadian Hospital Association, 1982). Comparable statistics from the United Kingdom are not available; however, a review of over 50 published and unpublished works in Great Britain from 1963–1975 in the area of patient nurse dependency suggests a significant level of interest and activity (Wilson-Barnett, 1978).

The proliferation of patient classification schemes is an interesting phenomenon and raises the question of what are the differences between schemes and ultimately which one is best. While there are probably thousands of classification or dependency instruments in various stages of implementation throughout North America and the United Kingdom, it is estimated that there are less than 25 uniquely different schemes. Moreover, it is likely that all of them contain at minimum the critical indicators of care suggested by Connor in 1969. The greater differences would appear to be in the area of forms design rather than in the basic tenets of classification. And it would seem that all, even proprietary developed schemes, undergo modifications during implementation within a given facility.

As suggested by Baar et al (1973), modifications are often introduced without sufficient regard for the fundamental structure of the original schemes. Despite the fact that the basic indicators of care suggested by Connor correlate highly with total care time provided, additional indicators are frequently added. While these indicators may not add to the statistical validity of the schemes there may be benefits associated with user acceptance, which after 20 years of solid experience with patient classification is now perceived as deserving of serious consideration. Since the successful application of patient classification almost always depends on the efforts of staff nurses, ignoring their opinions regarding the face validity of the schemes can and has led to disastrous results. Anyone seriously interested in the successful application of the schemes today places great emphasis on the education and involvement of staff nurses in the design, function and maintenance of the systems.

In spite of the widespread use of patient classification methods within hospitals, and in spite of the verbal and tacit approval of the methods by hospital funding bodies, there frequently remains a gap between the nurse staffing levels and patterns requested and those allocated. It would appear that the final argument regarding whose figures are 'right' will only be resolved when the incremental effect of specific staffing levels and patterns upon the quality of care can be demonstrated. And that definitely must await the development of valid, reliable, sensitive and comprehensive measures of the quality of nursing care.

Patient classification defined

Patient classification refers to the grouping or categorizing of patients according to the magnitude of the nursing care time required over a specified period of time. The primary purpose is to assist with the determination and allocation of nursing personnel resources. The categorization is based on a formal assessment of the patients' physical, psychosocial and teaching requirements. Patient classification, however, is distinct from patient assessmment. It is dependent upon and therefore comes *after* patient assessment, making use of the assessment information in a specific manner. Most patient classification methods are designed around 'critical indicators of care', that is, those characteristics of the patient found to correlate highly with direct nursing care time involvement. In the medical-surgical care areas, for example, research has demonstrated that the 'critical indicators of care' relate to such things as: the patient's degree of dependency in the areas of feeding, bathing, and ambulation; requirement for observation and special treatments; and requirement for psychosocial support and teaching (Aydelotte, 1973b). Thus, if one assesses patients according to the degree of nursing involvement with respect to each critical indicator, one can then group patients ranging from those requiring little nursing care time, to those requiring extensive nursing care time. Moreover, the 'critical indicators of care' are based on patient characteristics that are largely predictable. A nurse using professional judgement and knowledge of a patient's requirements for care, and the standards of care, is therefore able to predict the degree of patient dependency or requirement for some future time period, such as for the next day or next shift. Since patient classification methods were developed to aid in staffing, prior knowledge of patients' requirements for care is essential if one is to respond effectively with adequate staffing levels.

Before proceeding with a discussion of the application of patient classification it is important to clarify some of the terms commonly employed. Patient classification is the term used to refer to the instruments and methods employed to categorize or group patients. There are numerous possibilities for patient classification such as diagnosis and level of care; however, the generic term has become widespread in its reference to patients' requirements for *nursing* care. Occasionally, the term 'patient acuity' is used. It is generally understood however that acuity is inappropriate because of its association with the seriousness of a patient's condition or illness which does not necessarily correlate positively with requirements for *nursing* care. A second term, 'patient dependency' has also been used and is perhaps more descriptive of the true nature of most of the classification instruments, which are based on a dependence-independence continuum.

Clarification is also in order in reference to the phrase 'requirements for nursing care'. In many instances the terms 'needs' and/or 'demands' are substituted for 'requirements'. This practice reflects in part the multiplicity of the fields historically involved in the development of the instruments, such as economics, operations research, and management engineering. It also represents perhaps a measure of the carelessness with which we use the English language. We tread on dangerous ground when we assume we can identify the 'needs' of patients. Few would argue that a measurable sum of the care provided to patients is more a reflection of professional practices, policies and ritualism than actions in response to true 'needs'. Similarly the use of the word 'demand' is questionable since it is defined as an act of demanding or asking. Patients do not commonly 'demand' nursing care.

Finally, the term 'patient classification system' itself is occasionally subjected to debate. First is the question of the appropriateness of the word 'system', which entails debate that is beyond the scope of this chapter. Suffice it to say that what is generally intended by reference to 'system' is the classification instrument, the quantification or measure of nursing care time associated with each category, and the procedures employed to effect a staffing allocation method. Second is the question of the appropriateness of the term 'patient classification'. The argument here is that patient classification is limiting in that it is appropriate only for inpatient or contract clients and thus it represents only one approach to determining nurse staffing methods. More appropriate alternative terms would appear to be workload measurement systems, or simply staffing methods.

Approaches to patient classification

The many patient classification instruments and methods now available can be grouped as belonging to one of three major approaches: patient profiles, nursing task documents or critical indicators of care. Profile instruments are characterized by broad descriptions, generally in paragraph form, of the characteristics of a typical patient in each category. These types are also commonly referred to as 'prototype evaluations' (Abdellah & Levine, 1979). The actual characteristics of the patient are compared with those described in the profile and the patient to be classified is then designated to the category of care which most closely matches the profile or prototype description. Many of the early classification methods used the profile approach and the literature provides numerous examples of these and other classification approaches (Giovannetti, 1978; Abdellah & Levine, 1979).

The second approach, referred to as nursing task documents, employs a listing of the majority, if not all, of the direct nursing care tasks or activities that may be required by patients. Each task or activity is associated with a value representing the standard time required to carry out the activity, thus permitting an aggregated value of total care time for each patient. Following the determination of care requirements unique to each patient, categories are developed. For example, patients with a total point value of 1–20 are assigned to category I; similarly, a point value between 11 and 30 represents category II patients, and so on. Since an estimate of the unique time requirements of each patient is a product of the task approach, many have questioned why the final designation to a patient category is necessary. It would appear that the use of categorical information is helpful in making decisions affecting the overall determination and allocation of nursing resources. While it may be of interest to have knowledge of the individual care requirements of each patient, this level of precision has not been shown to be either necessary or helpful to staff determination and allocation. Measuring the unique care time requirements of individual patients represents a level of precision that has limited value in staffing decision given: (1) the value of cohesive and stable work groups; (2) the need for specialized skills; (3) the movement away from the assignment of tasks to various levels of personnel; and (4), the reality that the relative distribution of workload is not constant throughout the period of a shift. As noted by Giovannetti & Thiessen (1983), it is both practical and realistic to allocate staff on the basis of whole numbers and for complete shifts rather than fractions thereof.

The third approach, and by far the most common, is based on the critical indicators of care: the patient characteristics or elements of care found to correlate highly with the direct nursing care time required. Instruments which make use of critical indicators have also been described as 'factor evaluations'. Ratings on the individual patient characteristics or elements are combined to provide an overall rating which, when compared with a set of decision rules, identifies the appropriate care category. The instruments which make use of critical indicators of care differ primarily in the number of critical indicators used. In all cases, they contain as a minimum those indicators relating to activities of daily living described in the early work at the Johns Hopkins Hospital. The differences among the various critical indicator approaches are interesting in that there is little evidence to suggest that additional indicators add to the statistical validity of the process of classification.

A pilot study testing three different instruments based on critical indicators was conducted by Roehrl (1979) at the Medical Center Hospital of Vermont. The three procedures, which categorized patients into four categories, included one developed at the study hospital, the instrument reported by Hanson (1976) and subsequently modified by the U.S. Department of Health, Education and Welfare (1978), and an instrument developed at the University of Saskatchewan as reported by Giovannetti (1973). The findings of the 7-week pilot study, based on a sample of 779 medical and surgical patients, were that the highest correlation (.64) was achieved when the two outside classification instruments were compared with each other. When considering a two-classification difference, agreement between the two outside instruments increased to .99. The length of time needed to complete the classifications was also noted. The outside instruments took 12 and 15.5 minutes, respectively, to complete, while the finding of 27.5 minutes was reported for the study hospital's classification instrument. These findings, while limited to the study setting, represent one of the first attempts to determine the variability between different patient classification instruments in their distribution of patients among categories. For the study hospital, the findings suggested that the development of classification instruments unique to a facility was not necessary, and that more than one system could be used for interhospital comparability.

Debate continues among users and potential users of the many critical indicator approaches. As discussed by Hanson (1979), the selection of critical indicators is based not only on their contribution to the statistical validity of a classification instrument, but on their subjective desirability. Emotional support, teaching and specific treatments represent examples of indicators that, although not necessarily statistically predictive of patient requirements for care, are frequently retained because of their contribution to system credibility among nurses.

The end product of all three approaches to patient classification profiles, tasking documents and critical indicators, is the assignment of a patient to a specific care category. All three have been implemented in a variety of settings and apparently differ little in the educational requirements for implementation, time for application, and monitoring efforts. Following, is a brief description and illustration of two classification methods based on the critical indicator approach.

Figure 6.2 represents the *Medicus Patient Classification* Form appropriate for medical, surgical, obstetrical and pediatric patients. In all, 32 critical indicators are listed under the general headings of conditions, basic care and therapeutic needs. Patients are rated independently in each of the critical indicators and each indicator is represented by a unique weight. A ruler bearing the weights associated with each indicator is overlaid on the form. The process then requires the weight for the recorded indicators to be added. The total number of points then determines the patient's category. For example, patients with point values from 0–24 are designated as Category I; patients with point values from 25–40 are designated as Category II. In this manner, each category has boundaries which reflect the average time required to provide nursing care to that group of patients.

A second document essential to the correct interpretation of the Patient Classification Form contains the definitions of the indicators. While some of the indicators are self-explanatory others require interpretation. For example, the indicator 'special emotional needs' is to be checked if the patient and/or family exhibits one or more of the following behaviors: withdrawn, aggressive, anxious/demanding and expressing suicidal ideations. Further, the indicator is to be used for psychosocial disturbances which require specific nursing actions and therefore does not include normal amounts of comfort and support. Finally, the selection of this indicator must be accompanied by documentation of the patient's and/or family's behavior and nursing interventions.

It should be noted that virtually all classification instruments are accompanied by similar documents containing the meaning and interpretation of the critical indicators. In most instances a unique set of definitions is developed for each specialty area, thereby extending the use of one classification form to all units within a facility.

Since the Medicus Patient Classification Systems is represented by a proprietary establishment, the point value for each indicator, the methods used for establishing the point values and the resulting hours of care associated with each patient category are not in the public domain. For this reason, it is difficult to determine the manner and extent to which the unique characteristics of individual units within a facility are taken into consideration in determining staffing levels.

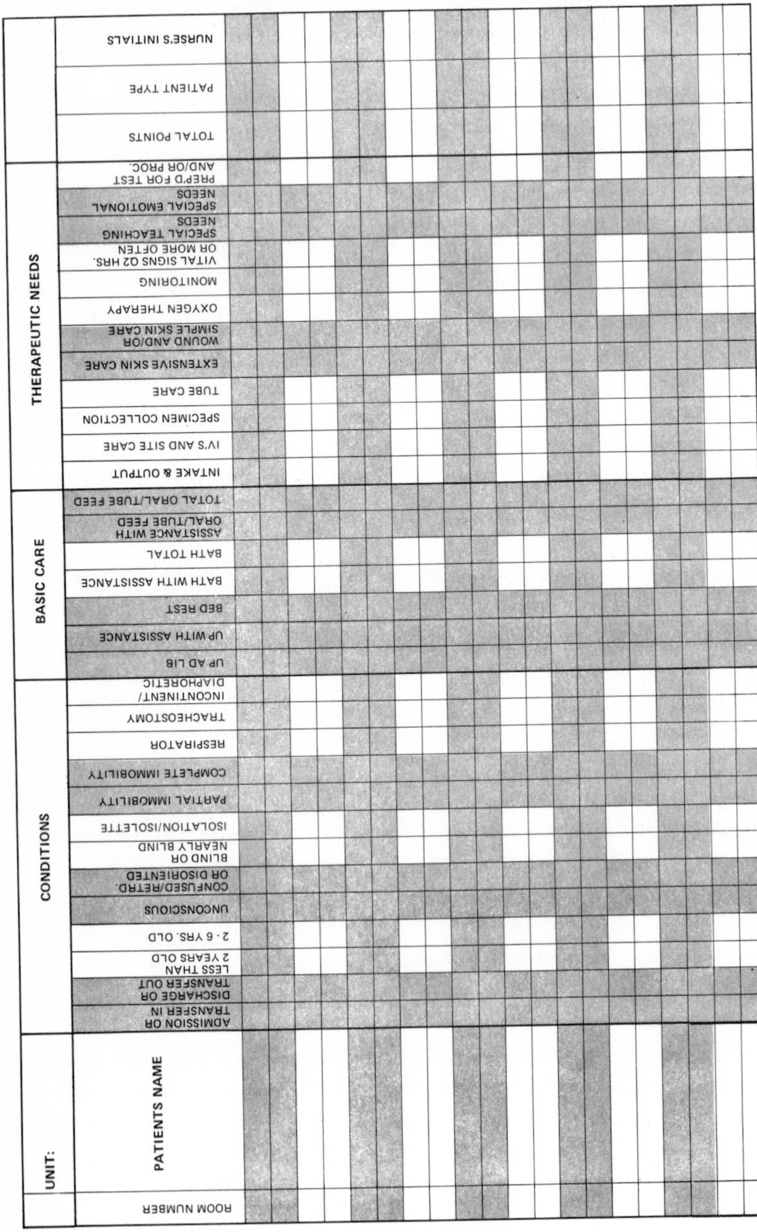

Fig. 6.2   Medicus classification from (reproduction by permission of Medicus, Canada)

As is the case with most other classification approaches, staffing patterns are not automatically determined. To determine staffing patterns, the Medicus Corporation takes into account, among others, the nursing department's philosophy and goals.

The Medicus System is operational in a variety of institutions and is designed for computerization. The corporation has developed an accompanying computerized management reporting system which includes productivity monitoring and staffing analysis.

The Public Health Service Patient Classification system, also representative of the critical indicator approach, assigns patients to one of four care categories ranging from class I, minimal, to class IV, intense (U.S. Department of Health, Education and Welfare, 1978). The format is a three-by-five inch card that may, if desired, be placed in the Kardex with the patients' medical orders and nursing care plan (Table 6.1). The format and the method of marking indicators were adapted from the classification instrument developed by Hanson (1976).

The classification form contains nine critical indicators and is accompanied by a set of definitions and guidelines specific to selecting the appropriate indicators. Two checkmarks and a value of .5 in the 'total' column are pre-printed on the form and serve as weighting factors to preclude most ties in patient classification. After the appropriate indicators are checked and added, the column containing the largest number of checks reveals the patient's category of care. The validity of the classification categories was determined by extensive direct patient care studies. Informal comparisons among this and other critical indicator approaches continue to reveal strong

Table 6.1   Public health service patient classification form

Patient class	I	II	III	IV
Activity Independent	( )			
Bath, partial assist		( )	( )	
Position, partial assist		( )	( )	
Position, complete assist			( )	( )
Diet, partial assist		( )	( )	
Diet, feed			( )	( )
i.v. Add. q 6 h or more or TKO		( )	( )	( )
Observe q 1–2 hr			( )	( )
Observe, Almost constant				( )
	(✓)	(✓)		
Total			.5	

Comments:

positive correlations in the assignment of patients to the four categories. Moreover, agreement has been high with nurses' professional judgement as to the ranking of patients from those requiring the least nursing care time to those requiring the most nursing care time.

One of the most important benefits of this approach is that it does not presume the amount of direct care time required to care for patients in any of the categories. The staffing method is accompanied by detailed instructions on conducting direct care studies that can be used to assign an 'average care time' value to each category. In this manner, the care times become unique to each nursing unit. Similarly, detailed instructions are provided for conducting nursing personnel activity studies to determine the amount of time nurses have to provide direct care.

The Public Health Service Patient Classification method has also been implemented widely and in a variety of settings. The computerized application of the system provides an array of management information reports.

Approaches to measuring nursing care time

Generally, when reference is made to patient classification methods, two distinct components are recognized: the classification procedure or instrument itself, and the hours of care or measure of nursing workload associated with each of the defined categories. The latter is often referred to as the 'quantification component' and is central to the primary purpose of patient classification, that is, the determination and allocation of nursing personnel resources. As previously noted, classification instruments are often transferable among similar groups of patients. This is due to the universality of the critical indicators of care. However, the hours of care or measure of the workload associated with each patient category are necessarily specific to the nursing unit and therefore are *not* readily transferable. For this reason, the implementation of patient classification requires that the user establish, for each unit and shift, the nursing care time required to deliver a specified standard of care to patients in each of the care categories.

Various methods for determining the time requirements for patients have been used. They can generally be considered as belonging to one of two approaches, those based on data collection, and those based on estimates. To date, there has been little effort directed toward identifying the effects of using different quantification methods, and the choice of method has rested largely with the preferences of either the agency involved in self-implementation, or the proprietary firms hired to develop and/or implement a system. What may perhaps be more important than the particular method selected, is the care and attention given to the process. Users are becoming more discriminating with the selection and approval of approaches, placing greater emphasis on issues relating to the reliability and validity of measurement techniques. The benefits of this are two-fold: greater credibility within the nursing profession, and greater acceptance by administration, both of which are central to establishing, maintaining and fine-tuning a successful staffing method. In the following sections both data collection methods and estimation procedures are discussed in greater detail.

Data collection methods

The two most commonly used methods for establishing nursing care time and involving formal data collection procedures are referred to as average care times and standard times. Average care times involve observing, either through continuous or intermittent (sampling) observational techniques, the amount of direct care time provided on each shift to a sample of patients within each patient care category. This procedure necessitates the implementation of a classification instrument *prior* to data collection. The average (or mean) direct care time provided on a shift basis to each category of patient can then be determined from the direct care observations. Since the critical indicators used in most classification procedures relate only to patients' requirements for direct nursing care, a second and concurrent observational study, usually based on sampling, is required to determine the time and nature of non-direct patient care activities performed by each category of nursing personnel. This aspect of data collection provides the information necessary to identify within a particular unit and for a particular shift the average amount of time available to nurses to provide direct care. A number of references describe the application of the average care time method (Giovannetti et al, 1973; Williams, 1977; U.S. Department of Health, Education and Welfare, 1978; Giovannetti, 1983).

The value of the average care time method is that it takes into account of non-direct patient care activities performed by nursing nursing practices, and environment. Furthermore, it provides an account of non-direct patient care performmed by nursing personnel. This information is critical in evaluating the need for change in the distribution of work among nurses, other health care and service personnel. A potential limitation of the average care time method is that it represents what 'is' vis-à-vis what 'ought to be'. The degree to which this distinction is significant depends upon the nursing resources available on the study units and the quality of nursing care being delivered. In some instances, when the resources are recognized as totally inadequate, the average direct care times have had to be augmented. Usually, this is done on the basis of professional judgment, consensus, and identified gaps in the delivery of care. It must also be noted that the opposite may be true.

The second data collection method is based on standard times and entails the assumption that most nursing care can be broken down into discrete tasks. The time required for each task or activity such as a bed bath, injection, dressing change, etc., can be determined and a 'standard' time for each task can therefore be directly measured or estimated. The inclusion of non-direct care time, and in some instances direct care components such as psychosocial support, are generally handled by employing a constant time factor for all patients. The standard time method is based on the assumption that the sum of the parts, i.e., tasks, is equal to the sum of the whole, i.e., total care time. The credibility of this assumption is diminished by the fact that nursing care is not delivered in a constant manner throughout the period of a shift. Studies identifying the relative distribution of workload have shown that there are periods throughout a shift when care activities are extensive as well as periods when care activities are few (Harman, 1972). Moreover, this method has been criticized for its task orientation which is often considered incongruent with professional nursing practice. Standard times are frequently employed with the nursing task approach to classification as used in the PRN 76 and GRASP systems (Chagnon et al, 1978; Meyer, 1978a, 1978b).

Estimation procedures

While not a quantitatively sound approach, simply estimating the time requirements has been an approach used when either the resources or desire for systematic data collection have been absent. Three estimation methods are described below and may be employed individually or in combination.

The first method involves a redistribution of the total average hours of care available or budgeted for a particular facility or unit to an 'average hours of care' figure for each category of patient on each unit and for each shift. The term 'estimate' comes from several sources. The redistribution is based on an estimate of the relationship or ratio of one patient category to the next, an estimate of the proportion of workload assigned to each shift, and an estimate of the average number of patients in each patient category.

A second approach is to ignore the total average hours of care per unit budgeted, and estimate on the basis of professional judgement and consensus the hours of care per unit that ought to be assigned to each patient category for each shift. Both these estimation methods have serious limitations in that they fail to provide quantifiable evidence that the hours of care assigned to each patient category and shift relate to the patients' requirements for care. Nevertheless, they can often be supported on the grounds that they are no less legitimate than the pre-established average hours of care per patient per day currently in use or budgeted.

The third method of estimation involves adopting or adapting the time requirement per patient category and shift used in another facility. This method can also entail serious limitations. However, as with the other two estimation methods, it may prove to be no less legitimate than the system it is replacing. If the 'borrowed time' requirements are based on sound data collection techniques, and the factors which affect staffing such as philosophy, standards of care, types of patients, extent of support services, physical layout and design are similar in the borrowing facility, then the results may be appropriate and useful. The challenge however is obtaining the appropriate data to make this judgment.

Applications of patient classification

As previously emphasized, the primary function of patient classification is to predict patients' requirements for care, the sole purpose being to provide assistance in determining the resources required to provide the care. The period of prediction is variable and depends to a large extent upon the stability of the care requirements of the patient population, and the practicality of altering staffing resources. In some instances classification is conducted once per shift; in others, once per 24 hour period; in still others, once per week or even less frequently.

The application of a patient classification method constitutes an attempt to bring more objectivity and structure to the process of determining the number of nursing personnel required to respond effectively to patients' requirements for care. This does not imply, however, that the methods should replace professional judgment on this matter. The responsible use of a patient classification system demands that the professional judgment of those charged with the provision of care be considered in making final staffing determinations. As illustrated in Figure 6.1, patients' requirements for care represent only one of many factors that influence staffing levels. In discussing the following applications, the assumption is made that the patient classification method employed has been determined to be both reliable and valid, and is used in conjunction with professional nursing judgment.

Baseline and variable staffing

As previously noted, patient classification methods have been developed to respond to the variable nature of patients' requirements for nursing care, usually accomplished in two stages, establishing an appropriate baseline staffing level and making use of variable staffing. In the first stage, patient classification information, aggregated over time, can be used to reveal the average range of care requirements on each shift and within each nursing unit. The findings aid in identifying the average staffing level or baseline that should be used for establishing schedules. This information is further used to determine the budget requirements for full-time equivalent nursing positions (FTEs). In many instances, problems arise from the fact that the baseline staffing levels are inappropriate; either too many or too few nurses are assigned to a nursing unit or to a particular shift.

The second stage involves the use of the shift or daily classification of patients to pinpoint the specific occasions on each nursing unit when more or less than the scheduled staff are required. Appropriate adjustments in staffing levels through variable staffing can then be made. This may involve the use of a float pool or the reassignment of nurses from one unit to another. It should be noted that an alternative to adjustment staffing levels can also be found through the selective assignment of patients to a nursing unit or patient transfers between units. Particularly in instances when occupancy levels are below 100%, thoughtful assignment of patients to units can greatly lessen the task of continually modifying staffing in an effort to attain appropriate levels.

Other applications

In addition to the above, several other applications may be realized. The equitable distribution of nursing assignments is a logical benefit of a patient classification system. Since each patient category represents a range of care time, individual nursing workload can be identified and efforts made to assure equalization. In some cases, equitable workloads have taken into consideration the education, experience and special skills of individual nurses. The daily hospital-wide nursing requirements or workload statistics can also be useful in monitoring workload trends for specific weekdays, weekends and statutory holidays. Similarly, monitoring workload trends over a period of months and years provides the basis for internal comparisons and long-range planning.

The systems may be used to monitor changes in medical and nursing practices, for example, the frequency of use of monitors, the impact of Swan-Ganz catheters, and the effect of special routines for evaluating postoperative patients. Based on objective problem analysis, patient classification is also helpful for establishing a priority setting plan to be used during periods of acute staff shortages. When staffing levels are inadequate, priority setting may allow the staff implicit approval to dispense with certain care activities and leave management with the security that unit based decisions made in emergency situations are appropriate. Similarly an evacuation or crisis plan can be developed utilizing patient classification information.

Patient classification information has also been used as a basis for selecting process criteria as part of a quality assurance program. For example, assessment criteria for patients requiring minimal nursing care could focus on basic nursing care and patient education, while assessment criteria for patients requiring extensive care may be more appropriately focused on complex nursing care and patient safety measures (Jelinek et al, 1974).

Finally, patient classification information can form the basis for determining the costs of providing nursing care. Whether or not the relative cost is passed on to the patient or the insurer, the determination permits separation between the actual cost of nursing care and the total service costs. The long-term benefits to nursing are likely to be the development of more rational budgets and the fostering of nursing accountability for fiscal management (Walker, 1982).

The above represents only a sampling of the potential applications of a well designed and implemented patient classification system. The benefits relate less to the particular method selected than to the level of understanding and commitment of the users. In all instances, however, assuring that the patient classification continues to be both reliable and valid is essential. For this reason ongoing maintenance mechanisms and monitoring strategies need to be established at the time of implementation.

Maintaining and monitoring patient classification

The development of staff orientation and continuing inservice educational programs, and the ongoing testing of reliability and validity are now considered as corequisites for patient classification.

Orientation and inservice education

Commitment to patient classification demands that the concept of patient classification be well understood. The significance of patient classification and its proper use should be a part of all orientation programs. The educational effort should be directed to all nursing staff who may be responsible for classifying patients; managerial personnel who may use classification information in management decisions; and other key personnel such as physicians and admitting personnel who may affect and/or be affected by the outcomes of the classification of patients. For staff nurses, an additional program for developing classification skills is required. The focus of this program is to establish an acceptable level of agreement or reliability among all nurse classifiers.

Inservice education is viewed as key to maintaining classification skills and should be undertaken as a long-term commitment. Periodic review of the patient classification system should be considered within the long-range educational programming. In addition, if user acceptability and/or reliability deteriorates, the educational program can serve to rejuvenate the system and increase reliability coefficients.

Reliability monitoring

Reliability refers to the consistency or the repeatability of the outcomes of the classification process. Since the classification of patients may be done by a number of different nurses on a number of different nursing units, it is important that all nurses be consistent. Patient classification methods do not come as 'reliable methods'. *Reliability must be established and maintained in each setting*. The consistency or accuracy with which a group of nurses can arrive at the same category of care for a group of patients must be determined at the time of implementation. It is also important that reliability be monitored periodically, and on an ongoing basis. This may be done by having an 'expert' classifier within the facility check all or a representative sample of the classifications done by the nurses on a particular unit. A simple percentage of agreement between the 'expert' and the nurse can then be used as the basis for determining the level of reliability. Generally, a percentage of agreement of 90% or over is considered acceptable. A percentage of agreement below 80% indicates the need for further education and skill development and/ or review of the classification decision rules.

In addition to the *categories* of care being consistent among different raters, the selection of the same critical *indicators* of care should also achieve a high level of consistency. In all cases, differences in ratings should be analyzed as to probable sources of error and corrective action should be taken.

Validity monitoring

Validity refers to the extent to which the classification instrument actually measures what it seeks to measure. It too must be established and maintained in each setting. Validity is generally assessed at the time of implementation. For example, the classification instrument is implemented and some form of direct care observation, standard time determinations, or professional consensus is used to determine the required hours of care for each of the patient categories. Validity, like reliability, must be monitored periodically. Changes in the practice of nursing care can occur to such an extent that the hours established for each patient category may no longer be appropriate; for example, the introduction of a standard that all patients on restraints will be monitored every 15 minutes. Similarly, institutional changes may dramatically alter non-direct care nursing time; for example, a reduction in the number of pharmacy support personnel. Each agency requires some means of monitoring such changes so corresponding changes can be made in the established hours of care.

One method of monitoring validity involves the implementation of a program to evaluate periodically the nursing staff's satisfaction with the established hours of care. This can be done by a questionnaire survey used to assess the opinions of the head or charge nurse regarding the adequacy of the staffing level on a particular shift. Although the evaluation is subjective, it has been found that the judgment of the professional nurse usually reflects the actual situation on the unit (Williams & Murphy, 1979). Professional nursing judgment is a crucial factor and should not be overlooked in the process of validity monitoring.

A second method of validity monitoring is the establishment of an expert panel of nurses to review periodically the hours of care per category for each unit. This can be done in conjunction with other monitoring processes and involves the expert opinion of knowledgeable nurses in the particular clinical area. Again, the reliance is on professional nursing judgement.

Finally, monitoring may entail the conduct of further direct and indirect care or standard time studies to verify the established hours of care. While this type of validation is more extensive, it may be required if major changes occur in either nursing practice or any of the factors known to affect staffing.

Only with the establishment of maintenance mechanisms and monitoring strategies will a patient classification system function effectively over time. Most of the problems faced in planning and implementing patient classification, as identified by Huckabay & Skonieczny (1981) in their survey of administrative nursing personnel, can be linked to the absence of appropriate education about patient classification and inadequate establishment of levels of reliability and validity among the instruments used. Historical experiences with reliability and validity provide little assurance of continued performance in this area. Reliability and validity must be re-established at the time of implementation and maintained throughout the life of the system.

RELATIONSHIP BETWEEN STAFFING AND QUALITY

As noted in the introduction, the assumption of a linkage between nursing resources and the quality of patient care is commonplace. Further, it is assumed that central to evaluation of patient care is evidence that patients' requirements for care have been both assessed and met. Patient classification methods provide the tools for delineating, in part, patients' requirements for nursing care and thus constitute a prerequisite to providing evidence that care requirements are met. Theoretically, at least, these notions appear to be reasonable. Providing scientific evidence, on the other hand, continues to present a real challenge to the profession. The supporting literature is sparse if not disappointing. The difficulties inherent in defining the relationship between staffing and quality stem from two sources.

First, there are the problems associated with establishing the validity of any assessment or means that attempts to define 'patients' requirements for nursing care'. Second, there is the two-fold problem of ferreting out the effects of nursing care from the care provided by others and developing reliable, valid and sensitive measures of the effects of that care.

Patient classification schemes have attempted to resolve the first problem by defining patients' time requirements for care. The extent of validity testing to support this has been primarily restricted to measurements of actual care given coupled with professional judgement as to the adequacy of the care. Moreover, it would appear that further validation is unlikely until we have a comprehensive set of reliable and valid measures of quality.

Assuming that patient classification systems will continue to form the basis of staffing methods, a cautionary note is in order. First, the existence of a patient classification system does not guarantee adequate staffing. For example, there is no assurance that the correct allocation of staff will be made. Second, there is no guarantee that the nurses will or can perform in the manner intended, i.e., having the skills or motivation to do so. This caution is necessary because it would appear that some believe that the implementation of a patient classification scheme will obliterate staffing 'problems'.

On the more positive side, we may view staffing methods that make use of patient classification, and the quality of patient care, as truly interrelated and contributory (Luke et al, 1983). While there should be little argument with the notion that the existence of a patient classification does not guarantee adequate staffing, a well developed and implemented system would seem to have benefits. The systems can be instrumental in developing staffing programs that attempt to match patients' requirements for care with nursing resources.

The two concepts are interrelated to the extent that resource allocation affects program performance. Further, it is assumed that effective utilization and allocation of nursing resources do have a positive influence on the quality of care. As suggested by Donabedian (1980), the allocation of time and other resources affect the performance of individual practitioners as well as the performance of programs.

On the basis of a survey of health care executives, Stinson & Ytterberg (1983) say they are perplexed if not dismayed by health care executives' reporting that while the economic crunch has effected a reduction in quantity of nursing staff, quality of care has not suffered. It is illogical to suggest that there is no relationship between the quantity of staffing and the quality of care.

SUMMARY AND CONCLUSION

The author has attempted to put forward a perspective on some of the issues germane to determining nurse staffing resources. A conceptual framework from which to view nurse staffing was provided to illustrate the numerous factors that need to be considered in responding to patients' requirements for nursing care. The multiplicity of factors affecting resource decisions have a concomitant bearing on the evaluation of that care. Indeed, the process is complex and in part contributes to the difficulty we have in linking staffing to quality.

A brief history of staffing methods was presented and, like all glimpses into history, can be helpful in understanding current practices. The staffing methods based on the classification of patients that emerged in the early 1960s have continued to provide the basis for current decisions affecting staff determination. While these methods continue to be useful, there appears to be a movement away from *total* reliance on the precise measurement of care to greater acceptance of accompanying professional nursing judgement in decisions affecting staff resources. This movement is appreciated in light of the array of factors that are believed to affect staffing.

The concept of patient classification, examples of classification instruments and the more commonly applied methods for quantifying nursing care time were discussed. While the process of patient classification is relatively simple, the instruments are not self-sustaining. Great care and attention is required to assure that they continue to be both reliable and valid. Moreover, the potential for misuse of classification methods is considered to be a real threat to users. Extensive educational programs are offered as a means of ensuring that the methods continue to function as intended.

In discussing the relationship between staffing and the quality of patient care, the sparcity of scientific evidence about the nature of the relationship was noted. While more positive demonstrations must await further developments in the measurement of quality, it is illogical to assume that there is no relationship between quantity and quality.

REFERENCES

Abdellah F G, Levine E 1979 Better patient care through nursing research, 2nd edn. Macmillan, New York
Aydelotte M K 1973a Staffing for high quality care. Hospitals 47 (Jan. 16): 58–65 +
Aydelotte M K 1973b Nurse staffing methodology: A review and critique of selected

literature. DHEW Publishing Number (NIH) 73–433. US Government Printing Office, Washington, D.C.

Baar A, Moores B, Rhys-Hearn C 1973 A review of the various methods of measuring the dependency of patients on nursing staff. International Journal of Nursing Studies 10(3): 195–203

Bernstein E 1953 A study of direct nursing care consumed by patients with varying degrees of illness. New York University, New York

Buchan I M 1979 Nurse staffing methodology in Canada. Canadian Nurses Association, Ottawa

Canadian Hospital Association 1982 Report on the patient classification systems survey. Unpublished report, Ottawa

Canadian Nurses Association 1966 Report on the project for the evaluation of the quality of nursing service. Canadian Nurses Association, Ottawa

Chagnon M, Audette L, Lebrun L, Tilquin C 1978 A patient classification system by level of nursing care requirements. Nursing Research 27(2): 107–113

Connor R J 1960 A hospital inpatient classification system (Doctoral dissertation). The Johns Hopskins University, Baltimore

Donabedian A 1980 The definition of quality and approaches to its assessment, vol 1. Explorations in quality assessment monitoring. Health Administration Press, Ann Arbor, Michigan

George F L, Kuehn R P 1955 Patterns of patient care. Macmillan, New York

Giovannetti P 1973 Measurement of patients' requirements of nursing services. In: Research on nurse staffing in hospitals. Report of the Conference. DHEW Publication No. (NIH) 73–434. US Government Printing Office, Washington, D.C.

Giovannetti P 1978 Patient classification systems in nursing: A description and analysis. DHEW Publication No. (HRA) 78-22, Hyattsville, Maryland (NTIS No. HRP 0500501)

Giovannetti P 1979 Understanding patient classification systems. Journal of Nursing Administration 9(2): 4–9

Giovannetti P 1983 Patient classification: Quantification methods (video tape). University of Calgary, Department of Communications Media, Calgary, Alberta

Giovannetti P, Thiessen M 1983 Patient classification for nurse staffing: Criteria for selection and implementation. Alberta Association of Registered Nurses, Edmonton, Alberta

Giovannetti P, McKague L, Bicknell P 1973 The development, implementation and evaluation of a workload index for Holy Family Hospital. Hospital System Study Group, Saskatoon, Saskatchewan

Hanson R L 1976 Predicting nurse staffing needs to meet patient needs. Washington State Journal of Nursing, Summer-Fall 7–11

Hanson R L 1979 Issues and methodological problems in nurse staffing research. Communicating Nursing Research, vol. 12 WICHE, 51–56

Harman R 1972 Getting more out of personnel utilization studies. Hospital Administration in Canada 14(Oct): 80–83

Huckabay L M, Skonieczny R 1981 Patient classification systems: The problems faced. Nursing and Health Care February: 89–102

Jelinek R C, Dennis L C 1976 A review and evaluation of nursing productivity. DHEW Publication No. (HRA) 77–15, Bethesda, Maryland

Jelinek R C, Haussmann R K D, Hegyvary S T, Newman J F Jr 1974 A methdlogy for monitoring quality of nursing care. DHEW Publication No. (HRA) 76–25. US Government Printing Office, Washington, D.C.

Luke R D, Krueger J C, Modrow R E (eds) 1983 Organization and change in health care quality assurance. Aspen, Rockville, Maryland, ch 16, p 243–252

Meyer D 1978a GRASP: A patient information and workload management system. M.C.S., Morganton, N.C.

Meyer D 1978b Workload management system ensures stable nurse-patient ratio. Hospitals 52(Mar. 1): 81–82+

Nadler G, Sahney V 1969 A descriptive model of nursing care. American Journal of Nursing 69: 336–341

National League of Nursing Education 1937 A study of nursing services in 50 selected hospitals. The United Hospital Fund of New York, New York

National League of Nursing Education 1948 A study of nursing service in one children's and 21 general hospitals. National League of Nursing Education, New York

National League of Nursing Education 1949 Criteria for assignment of the nursing aide. American Journal of Nursing 49: 311–314

Roehrl P K 1979 Patient classification: A pilot test. Supervisor Nurse 10(2): 21–22+

Stinson S M, Ytterberg L 1983 University of Alberta, personal communication

US Department of Health, Education and Welfare 1978 Methods for studying nursing staffing in a patient unit. DHEW Publication No. (HRS) 78–3, Hyattsville, Maryland

Walker D D 1982 The cost of nursing care in hospitals. In: Aiken L H (ed) Nursing in the 1980s: crises, opportunities, challenges. Lippincott, Toronto

Williams M A 1977 Quantification of direct nursing care activities. Journal of Nursing Administration 7(Oct): 15–18

Williams M A, Murphy L N 1979 Subjective and objective measures of staffing adequacy. Journal of Nursing Administration 9(11): 21–29

Wilson-Barnett J 1978 A review of patient-nurse dependency studies. Department of Health and Social Security. Unpublished report, London

Wolfe H, Young J P 1965a Staffing the nursing unit, part 1, controlled variable staffing. Nursing Research 14(3): 236–243

Wolfe H, Young J P 1965b Staffing the nursing unit, part 2, the multiple assignment technique. Nursing Research 14(4): 299–303

Wright M 1954 The improvement of patient care. G.P. Putnam, New York

Young J P, Giovannetti P, Lewison D, Thoms M 1981 Factors affecting nurse staffing in acute care hospitals: A review and critique of the literature. US Department of HEW, Contract No. (HRA) 232-78-0150 (NTIS No. HRP 0501801), Hyattsville, Maryland

Answering the question with electronic instrumentation

Research and practice are interdependent. Research, without links to professional practice, can become sterile and irrelevant. Professional practice, without links to research, can become routine, repetitive, unscientific and stultified.   The two must go together. Increasingly, research into the delivery of nursing care should contribute to quality of care as well as to the best use of nursing manpower. To improve practice, we must do more than just evaluate the care of a given patient. We must be able to document the extent to which the outcomes were influenced by the clinical interventions used. If patients improve at least in part because of the clinical approach used, the implications for practice are totally different from what they would be if the patient improves despite the approach. Nurses therefore have a professional responsibility to analyze critically both their practice and itts effects on care. This chapter will outline the steps of the research process. Specific emphasis on the use of electronic instrumentation as part of the data-collection tool will be given. Examples of studies using electronic data-collecting tools to illustrate the rationale for the methodology and the problems and joys of the methodology, data collection and analysis phases will be described.

THE RESEARCH PROCESS

What is nursing? What is a good nurse? What constitutes nursing excellence? The process of both research and quality appraisal is characterized by a process of systematic enquiry. The sequence of steps of the quality assurance model, depending upon the model chosen, might have differing terminology but the component ideas usually reflect the following stages:

identification of an agreement upon values;
choice of criteria;

establishment of standards for outcome, process and structure;
ratification of criteria and standards;
evaluation of current levels of nursing practice against the ratified
standards;
identification of and analysis of factors contributing to evaluation
results;
selection of appropriate actions to maintain or improve care;
implementation of selected actions;
re-evaluation (Treguna, 1977).

In comparison, the main steps of the research process are:

identification of a researchable problem;
assessment of available resources such as time, money, expertise
and available literature;
design of the study;
development of data-gathering tools;
data collection;
data analysis and interpretation;
preparation of the research report.

The development of a research plan is essentially a process of
moving back and forth through the different steps and of continuously expanding and polishing. This process of moving back and
forth occurs in the appraisal process. It progresses to implementation of further actions and re-evaluation which may or may not be
part of a particular research study.

To plan a research project generally involves translating a hunch
or rather vague interest in or curiosity about a particular problem
area into a specific problem which can be researched. Research ideas
are often generated from gaps in knowledge that exist between
theory and application or from hunches about relationships between
certain events or variables. The more specific, refined or well-stated
the question, the easier it is to determine the method for answering
the question. Once relevant and specific questions are proposed, an
investigator may speculate on the possible answers to the question.
This is the process of hypotheses generation.

The review of resources necessitates a critical survey of the available literature together with an assessment of resources. This review
proceeds over time with further refinement of the problem and
purpose. Nursing is not a pure discipline; it is a 'science-of-practice'
discipline. We use theories from basic sciences in our deliberations.
The review assists in building a theoretical framework. Not only is

one interested in results of previous research, but also in the research approaches and methods tried, the data-collection instruments developed, the techniques of data analysis and the successful and unsucessful elements.

The selection of the research design can be done with greater insight following the review of the previous research. Historical, survey (descriptive) and experimental are major approaches. The historical approach looks at what has happened in the past to see current conditions and problems with a deeper and fuller understanding. The survey (descriptive) approach sheds light on current problems by describing and understanding current conditions. The experimental approach usually seeks to explain what happens to A when B occurs. The decision to employ one type of design or another is often not a matter of free choice but depends upon the problem selected for study.

The research methodology refers to the procedures and the sequence of procedures an investigator uses to conduct the study. This concerns how the study population will be selected and how the data will be collected. It involves choosing an appropriate data-gathering technique and selecting or developing the data-gathering instruments. An instrument refers to the tool or equipment used for collecting data.

After a problem is identified, the important questions for the researcher are what to measure and how to measure it. In looking at what to measure, one is looking for indicators that tell us what we want to know. There are physiological indicators such as measurable signs and symptoms and psychosocial indicators such as marital status, home commitments, family structure. Some indicators are quite clear and precise while others are at various stages of development. For example, a person's temperature is a good measure of the presence of fever. Indicators of patient satisfaction are less well developed. Multiple indicators of the same concept might in combination reflect a fuller and more accurate picture. Anxiety is indexed in a stronger way if a combination of psychosocial and physiological indicators are used. Some variables that require measurement have no current indices, particularly those specific to nursing interventions. The availability of sound measures will advance the testing of theoretical formulations underlying nursing practice. Improvements in this area will enhance our ability to add to the existing body of nursing knowledge, increase our ability to assess current practice, and enhance our ability to develop nursing interventions and programs to influence positively health care and client behaviour

(Ventura et al, 1981).

When data can be quantified, the ability to apply meaningful tools of statistical analysis is possible. Not all research deals with quantified data. Anthropological research depends heavily upon verbal descriptions and other types of narrative data rather than on numerical data. Often the nature of the variables will dictate the method used. One of the critical, if not the most critical, aspects of evaluation and appraisal of reported research is the quality of the research instruments. No piece of research can be better than the instruments used to gather the data. If these lack any of the basic attributes of good instruments, then the research data will be affected; sometimes this effect is serious enough to negate entire projects. If the research instruments possess all of the desired attributes, then the potential for good research is increased. These desired attributes include validity, reliability, sensitivity, generalizability, objectivity, and appropriateness (Fox, 1982).

Validity refers to a data-gathering tool or test actually testing what it is supposed to test. To use a thermometer to test temperature is valid but to use a thermometer to reflect diabetic control is not valid. A test does not have to be consistent over time to be valid, for example a pulse varies with a variety of factors. However, if the tool is not valid then one can see the serious flaws in the results and conclusions which could occur. The instrument should be constructed to contain components that get to the heart of the problem and that tap the vital elements of the question being studied. The instrument must be based on the theoretical framework selected for study. Reliability refers to the accuracy of a measuring instrument. It should provide the same data every time the subject uses the instrument. There are two basic sources of inaccuracy: one is error in the tool itself and the other is inconsistency between different individuals who are taking readings. 'An instrument may be highly reliable without being valid, but it cannot be valid without being reliable' (Treece & Treece, 1977).

Sensitivity refers to the ability of an instrument to make the discriminations required for the research problem. Are the scales for different variables of sufficient sensitivity to discriminate and be selective? Can they detect the significant differences among study subjects? The instrument should be able to pick up changes and differences in the areas studied, for example, changes in blood glucose levels should be detected through the accurate performance of particular tests.

Generalizability refers to the ability of the instrument to be used in various settings and with various clients and still be valid and reliable. Conclusions of a study from one setting cannot be used lock-stock-and-barrel in another setting without further examination. Appropriateness refers to the extent to which the respondent group can meet the demands exposed by the instruments. An instrument should then be appropriate for the clientele selected, for example, people who have difficulty with language should not be asked to read and write questionnaires.

Objectivity refers to the extent to which the data obtained are a function of what is being measured. If a person's eyes are closed can one assume that the individual is sleeping?

The need to estimate the validity and reliability of instruments under less controllable field conditions such as in the patient's hospital area, as well as under highly controlled laboratory settings requires techniques and strategies. Stinson (1973) cited several reasons why some nursing research is not valid and reliable. She says that much nursing research is often exploratory and the findings are not definite and the research design of some studies may be weak thereby leading to untrustworthy findings and conclusions. One must be careful about generalization of findings as there may be too many other factors not accounted for in the research study which make application from one area to another unsound. Investigators may also have used inappropriate statistical methods to analyze their data. She also states that conclusions are sometimes unsound based on the findings of that study.

The feasibility of using the instrument is an important consideration. Instruments vary in their ease of application. Some require more time, money, and personnel than others and if resources are limited, this can play an important part in the choice of instruments. The adequacy of the instrument can best be assessed by pretesting it before it is applied to the subjects in the actual study. This is a 'dry run'. Sufficient time should be allowed for the pretest to analyze fully the results so as to make all necessary changes in the data-collecting procedures.

Electronic devices as data-collecting tools

The selection of quality data-collecting instruments is a significant step in the research process. Many instruments are now available that can provide quantitative measurements on a self-recording basis. A variety of devices known as electronic patient monitoring apparatus are becoming available as tools for diagnosis and therapy;

these can also be used as instruments for research. The instruments remove one possible source of bias. Measurements can also be provided in a highly precise and sensitive nature. The use of electronic instrumentation thereby supplements, refines measurement, increases objectivity and permits accurate recall.

Some basic information about the facts and use of electronic instrumentation includes knowledge about the workings. Sensors are devices such as electrodes and thermometers designed to note or sense changes in physiological indicators. The sensor receives the energy input whether it be from electricity, heat, pressure, force, displacement or light. This energy is then changed into a signal which represents the change in energy and is usually in the form of a wave of electrical pulsation. This is then changed to a readout for display. The data can be processed to be a form of numerical display, a printout, a wave form on graph paper or oscilloscope, or a sound on tape. Note that the wave form is a representation of the phenomenon rather than the energy event received. Because it is a representation, one must consider the accuracy, fidelity and precision of the representation of the real event. Accuracy represents freedom from error or a measure of closeness to the true value of the event. Fidelity, by contrast, signifies replicability of the input or accuracy of the reproduction. Precision describes the fineness of exactitude of measurement. With high precision, the same event will give the same measurement under identical conditions repeatedly (Geddes & Baker, 1969).

When using electronic instrumentation, noise, systems isolation and time resolution need to be considered (Abbey, 1975). Noise is unwanted spurious signals of any type, such as static and artifact. Sources of electrical interference can be reduced in the design construction of the electronic equipment by including shielding to dampen those sources of noise. When measuring cardiograms for example, contact of another person with the individual being tested will distort the readout. Use of more than one electronic device on a patient, can affect measurement due to distortion. Calibration of the equipment can offset the effects of noise and system interference. Calibration is setting of the machine to a known standard, so that for example, zero level is at zero level and the units increase as required. This thereby provides a baseline. Over time the influence of a variety of factors can affect the precision of the reading. Gross patient movement, muscle tremor, power surges, changes in the resistance of wires connecting the processor to the transmitter can lead to loss of calibration (Jacobs, 1978). It is therefore advantageous to calibrate the equipment at the beginning and the end of each session since this promotes accuracy and precision.

Time resolution refers to setting the device so that the time-base input which is displayed or recorded is compatible with the event. The shape of the wave with its pattern and characteristics should not be distorted by being run too fast or too slow. It is important to know what the wave form should look like in order to catch problems as early as possible.

Major advantages to the use of electronic instrumentation include replicability of technique with standardization and accuracy of measurement as well as the objectivity of the recording. The use of these devices, however, can also drive a researcher rather silly.

The use of electronic devices usually requires that they have been checked by the hospital's bioengineering department for their compliance to electrical and safety standards. Standards for hospitals are higher than general household appliances in terms of allowable current leakage, design and other matters. However, most pieces of equipment designed for medical purposes will have been designed with hospital specifications in mind.

Possibility of equipment breakdown exists. The sophistication of many of the instruments requires considerable expertise on the part of the user. It is therefore advisable to have technical assistance available. In one's quest for research funding this need to consider possible repair/replacement requirements is important. As well, when considering funding, one must anticipate that the data collection period usually takes longer than originally planned. This is often true of most research but doing clinical research coupled with the use of instrumentation seems to add more uncontrollable elements to the pie. The significance of a 'dry run' cannot be underestimated to determine the 'bugs' and help work them out. In proceeding with the data collection plan, the use of checklists and double checks might help prevent and/or reduce problem areas.

The peculiarities of locating oneself and one's equipment in the appropriate place can be challenging. A large piece of equipment can be obtrusive and in itself can affect the events thereby affecting the results. Setting up cables or electrical wiring can also present problems in going over, under, and around obstacles which one might not have thought existed beforehand.

If the equipment will continue to operate when an observer is not present, there needs to be ways the subject can indicate discontinuation of their desire to participate in the project at any time. Something which might be rather small but can end up being rather a large problem is consideration of how the recording material will sit and be collected when the investigator is not present. The recording paper might be consumed at the rate of one meter per minute and unless that has been considered, one might find it all over the floor or used up during the interval or the recording pens might have required refill or replacement during that time.

Since the amount of data generated can be horrendous, the researcher needs to have considered this possibility in the planning phase or she/he might be 'snowed under' with all the paper. A plan for the analysis is essential otherwise one might have an unmanageable problem. The time to consider those issues is before the research data have been collected. If necessary, revisions can be made.

The data collecting phase begins the implementation stages of the research. Supposedly the planning and decision-making stages have been completed earlier, but in practice, this rarely occurs. Decisions may have to be made at the data collecting point if unexpected things happen. Such decisions are made with the aim of realizing as fully as possible the original data-gathering and data-analysis plan.

The data-analysis phase is largely a matter of putting a previously established plan into effect. It involves examination of the raw data, categorization, summarization, analysis and interpretation of facts and observations. Examination of the findings in light of the objectives or hypotheses is done. The research report recapitulates the stages of the process, explains decisions that have been made and their rationale, and presents the results of the research in the context of the specific research problem.

ILLUSTRATIVE RESEARCH STUDIES

Some major points in relation to selection and use of tools to measure the research question have been reviewed. These will be illustrated in the following brief report on studies in which the author has been involved. Additional information about each study can be obtained from the full published reports listed in the references.

Sleep study (Hilton, 1976)

Sleep is something familiar to us all. 'We sleep around 20 years in a lifetime, a considerable period to lie almost immobile, remote from the working world, rising and falling in waves of emotional experiences that are almost forgotten' (Luce, 1971). Inability to obtain sufficient sleep, a basic and vital need, is a common complaint of hospitalized patients. Because of the critical nature of the patient's illness in intensive care units and essential need for continuous monitoring and care, the normal sleep patterns and cycles are highly susceptible to interruptions and changes. What sleep patterns do those patients demonstrate while in intensive care and what effects them? This was the problem identified for one of my studies. The purpose of the study was to document the quantity and quality of sleep and to identify the factors that disturbed the sleep of selected patients in a respiratory intensive care unit (ICU). The purpose included documentation of behaviour which might be related to sleep deprivation and identification of factors which the patients in the respiratory ICU perceived as interfering with rest and sleep. Sleep literature including studies conducted both in controlled settings and patient care settings was reviewed. Nurses must often utilize theory developed from the other sciences. Much of the theory on sleep was developed in controlled sleep laboratories. Sleep is characterized by rhythms and cycles and the average normal night's sleep consists of four to five sleep cycles, with each being 90–120 minutes long. Four sleep stages and the so-called dream sleep in which there is rapid eye movement (REM) generally occur in each cycle in a predictable pattern; each is characterized by specific brain wave patterns, eye movement and skeletal muscle tension. There is an orderly progression of stages: 1, 2, 3, 4, 3, 2, and REM sleep. Predictable percentages of different stages in a night's sleep have been determined for various age groups. Sleep needs also vary from as little as 3–4 hours a night to an average of 7–7½ hours.

Sleep deprivation produces lower levels of arousal, increases in stages 3 and 4, decreases in stage 1, and decreases in the time taken to go to sleep. Following either stage 4 deprivation, there is a significant increase in stage 4 sleep even at the expense of other sleep stages. In severely dream-deprived subjects, the REM stages are found to succeed wakefulness rather than deep sleep. General sleep deprivation effects include lassitude, decreased mobility, sensory changes, visual and auditory illusions, hallucinations and disorientation.

This literature review gave the researcher direction in terms of design and methodology. A descriptive approach was selected due to the scarcity of data about sleep of patients in hospital.

An electroencephalographic recorder was used to measure brain wave patterns (electroencephalogram — EEG), eye movements (electrooculogram — EOG), and skeletal muscle tension (electromyogram — EMG) on ten patients in a respiratory ICU. Standardized

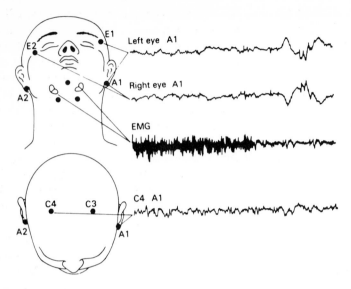

Fig. 7.1 Placement of EEG, EMG, and EOG electrodes for recording sleep stages (reproduced with permission from Rechtschaffen & Kales, 1968)

nomenclature and methods for recording and scoring sleep recordings developed by the Association for the Psychophysiological Study of Sleep (APSS) were used as a source of technique (Rechtschaffen & Kales, 1968). The polygraphic tracings were made at a paper speed of 15 millimetres per second on an eight channel Mingograff electroencephalographic machine. Two channels of EEG, two channels of EOG and one channel of EMG were used. Needle electrodes were used for scalp placement, ear clips and surface electrodes for eye, ear and chin areas. Figure 7.1 illustrates the placement of these electrodes. Subjects were each monitored continuously for 48 hours. Operational definitions had been prepared for sleep, normal sleep cycle, behaviour related to sleep deprivation and definite sleep-disturbing factor. Factors that might be interpreted as sleep-disturbing to the patient or that might have prevented patients from resting or sleeping were recorded directly on the polygraphic recording to co-ordinate timing for later analysis and cause-effect possibilities. Observations of unusual behaviour which might have been related to sleep deprivation were also recorded. Approximately one week following transfer from the ICU, each patient was interviewed to ascertain his normal sleep patterns; to identify factors which he or she believed prevented the obtaining of rest and sleep; and to record the recall and perceptions of unexplained experiences in the ICU. Other sources of patient data were used to complete patient profiles and to determine whether drugs for analgesia or sedation had been administered during the study period. A pretest was performed on one patient. The data was then collected from ten subjects who met specific criteria. As indicated, analysis of the records was carried out according to accepted guidelines. Records were checked for reliability of scoring by a neurosurgeon involved in sleep studies.

The findings indicate patients had less total sleep time than normal and percentage of time in the various sleep stages was not that of the normal cycle. Total sleep time ranged from 6 minutes to 13.3 hours during a 24 hour period. Only 50–60% of the sleep occurred during the night period as contrasted to almost 100% of that they normally experienced at home. Stage 1 predominated to the deprivation of all other stages; stage 4 (deep sleep) and REM (dream) sleep were almost non-existent. Quality sleep is dependent upon the orderly progression of stages and quantity of time in each stage. Poor quality sleep was evident in all subjects. No complete sleep cycles were experienced. Subjects had interrupted sleep periods throughout the 48 continuous hours, the longest sleep periods ranging from 42–96 minutes.

Factors that definitely disturbed the patients' sleep, as evidenced by the change in the polygraph recording towards wakefulness, were mainly noises created by staff members, for example talking to each other (22%), and environmental noises (20.8%) such as squeaky doors, paper rattling and background noise. Assessment (16.7%) and therapeutic procedures (12.5%) were the next most influential sleep-disturbers. The ranking of occurrence of possible sleep-disturbing factors was similar to that of the definite sleep-disturbing factors. The frequency of occurrence did not differ significantly from day to night. An inverse relationship existed between the number of factors occurring and the quantity of sleep experienced by the patient during the night period, that is, patients experienced more sleep if the possible sleep-disturbing factors decreased. The number and duration of possible disturbing factors therefore influenced adversely the quantity and quality of sleep. Two thirds of the subjects experienced behaviour related to sleep deprivation, such as nightmares, restlessness and hallucinations. Two patients were later unable to recall their experience in the ICU.

The implications of the study are that more judgement is needed in deciding when patients should be disturbed; many sources of stimulation which cause disturbance could be eliminated, controlled, or co-ordinated; and more methods of relieving patient anxiety and promoting physical relaxation could be employed. Consideration and follow-through of these factors together with a higher priority for patients' need for sleep could promote a more restful atmosphere conducive to better quantity and quality sleep.

Hypotheses generated from the research included:

1. If the possible sleep-disturbing factors are reduced, co-ordinated, eliminated or minimized, the patient obtains better quantity and quality of sleep;
2. Patients who obtain quality and quantity sleep experience fewer complications and recover more rapidly than do patients who are deprived of quantity sleep;
3. Relationships exist between the quantity and quality of sleep experienced and the development of signs and symptoms which are part of the intensive care syndrome.

Since only one respiratory ICU was used, the sample was small and select, and this investigation was descriptive in nature, no generalizations could be made. However, it could be concluded these patients had difficulty meeting their normal sleep needs because of frequent interruptions and possible sleep-disturbing factors. Patients were deprived of total sleep time, normal distribution of sleep stages and quality sleep. Behavioural changes related to sleep deprivation were observed. Most disturbance was caused by noise, the source of which was staff communication, environment, equipment and the other patient and his care.

Why I did what I did

There was a dearth of previous research on sleep done in patient clinical areas and most of this depended upon subjective indicators of sleep (such as eyes open/closed, depth of breathing or patient comments or periods of non interruption). Since these subjective measures are not reliable as measures of the presence/absence of sleep, nor do they indicate sleep staging, an objective method with electronic instrumentation was selected. The specific staging could be determined by recording EEG, EOG and EMG. The reliability and validity of the recording and scoring method by APSS indicated assurance in the data-collection and analysis process. As well, occurrence of an event could be evaluated in terms of whether it had any effect of waking the patient or lightening the stage of sleep.

Since sleep could occur at any time, a continuous period of monitoring was selected. The occurrence of sleep deprivation signs would be more likely to happen in a study lasting 48 hours rather than a shorter interval.

A pretest was performed. The investigator also recorded her own sleep by the same methods in the laboratory for one night to determine whether the patient was being exposed to any additional discomforts by participating in the study. It was concluded that the patient was not being exposed to additional discomforts.

Implementation of the research plan in the data-collection phase brought to light many of the unexpected problems referred to earlier. Noise, that is, the interference in the signal came about with one patient who was on a rocking bed and therefore had to be eliminated from the sample. To prevent possible problems in terms of coupling and any possibility of electrical interference, the sampling criteria had to exclude individuals with pacemakers.

The size of the EEG machine made the already tight quarters even more confining to patient and staff. The staff were receptive to the researcher and her equipment. Both to gain better skill in the area herself as well as to promote rapport with the health professionals, the researcher worked as a general staff nurse in the area for 5 months prior to the data collection. Perhaps this helped pave the way. At times the observer was not able to remain continuously with the patient because of other commitments and on return might find electrodes displaced and recording a straight line. Patients went to surgery or required transfer out of the unit for other procedures; this required discontinuation of the research for periods of time thus contributing to incomplete information on some subjects.

Funding is needed in order to have some control over the observers: what they do and when they do it. This study relied upon another student and some assistance from a department to do the day and part of the evening observations. The investigator started the patients on the study after obtaining the required consents from doctor and patient and did the remainder of the evening and all night observations. The sleep patterns of the researcher would have made an interesting study in themselves by the end of the data-collection period! One patient per week was studied and the data-collection phase took 3 months.

At a paper speed of 15 millimeters per second, one meter of data was produced every minute! One had to think ahead as to the time the paper would run out or the pens would run dry. One night at 0230 the pens ran dry and the supplies were locked inaccessibly inside a department. Occasionally, the recording paper might not fold and fall neatly into the paper catcher. Since this occurs quite silently one gets rather a start when one walks in to discover paper collecting beside the machine on the floor. The analysis phase of the recordings progressed simultaneously with the data collection phase using the established criteria. However, 48 continuous hours of recording on each subject provide materials for quite a 'lengthy' task. Fortunately, the reliability checks on the analysis proved positive. One knew then that both the data and the analysis of that data were accurate and thereby findings for those individuals was reflected. A nice way to end the study. Imagine my feelings if these had proved inaccurate!

Noise study (Hilton, 1982)

The indication that noise was the chief disturber in the sleep study generated further questions about noise levels in patient care areas. This led to a study designed to assess noise in various patient care areas of hospitals, including both intensive care and ward areas. Although deafening hospital noise has not been reported, there is suggestion that noise levels are high enough to cause adverse effects on patients. The specific objectives of the study were:

— to determine the level of noise dB(A) generated by mechanical equipment, nursing measures, and conversation and movement of personnel;
— to identify aspects of the acute care environment that require modification to limit noise in the immediate vicinity of the patient;
— to identify measures that help to reduce disturbing noises;
— to determine if there is the potential for noise levels to interfere with patients' sleep;
— to determine patients' perceptions of noise and its effects on them.

The literature revealed important information. Noise is technically defined as any unwanted sound having properties of frequency and intensity. Frequency represents the pitch of the sound and is expressed in cycles per second (cps). Intensity represents the loudness of sound and is expressed in decibel (dB) units. Sound levels in dB are calculated on a logarithmic basis, an increase of 10 dB representing a ten-fold increase in acoustic energy. The human ear also works logarithmically; therefore, perception of the noise increase of each 10dB is perceived as approximately a doubling of loudness. A weighted 'A' scale approximates the frequency of the human ear by placing emphasis on the frequency range of 1000 to 6000 cps, thus measurement uses a unit called dB(A). Measurements of sound level may be averaged over two distinctly different points of time. Variable sounds can be measured with a longer average time over periods of hours if necessary, and are expressed in terms of the equivalent continuous sound pressure level (Leq). This convenient measure of average noise exposure using the A-weighting correlates reasonably well with many human responses to noise and is recommended for general use.

Noise is associated with a host of bodily and mental changes. Elevated noise levels can have disturbing physiological effects, such as increased adrenalin production, peripheral and coronary artery vasoconstriction, respiratory rhythm alteration, decreased digestive secretion and motility, enhanced pain perception, altered perceptual processes, fatigue, impaired judgement and irritability and increased annoyance and stress. The threshold of tolerable sound is considerably lower for ill persons than those enjoying good health. There are also marked inter-individual differences in behavioural awakening thresholds. Noise studies in clinical settings used various approaches to data collection.

Recommendations in hospital inpatient areas include ambient noise levels of 40 dB(A) as acceptable, rising to 50–55 during intermittent noises, depending on the number of occurrences (Walker, 1978). Examples of noise levels from the threshold of hearing to the threshold of pain (0 to 140 dB(A)) are 60 for ordinary speech, 70 for street traffic, 20 for rustling leaves, 34 for soft whisper at 5 feet, 92 for heavy city traffic, 70 for noisy office and 80 for noisy factory.

A small, highly sensitive microphone on the end of a cable was suspended from the ceiling or attached out from the wall at the head of the patient's bed about 6 cm from the wall. The cable connected the microphone to the noise level meter which was in turn wired to the noise level recorder (Fig. 7.2). The range was set at 40–90 dB(A) and the calibration done accordingly. The paper speed ran at 1 mm/second when the observer was present. When not present it was set to run at .3 mm/second. In order to measure the equivalent continuous sound pressure level (Leq) a small sound meter with a data storage unit designed to measure Leq was suspended near the patient's headboard. This was later connected to a noise analyzer which analyzed the data for selected intervals (minute, 15 minute, hour, 24 hour average). This was printed out in both graph form

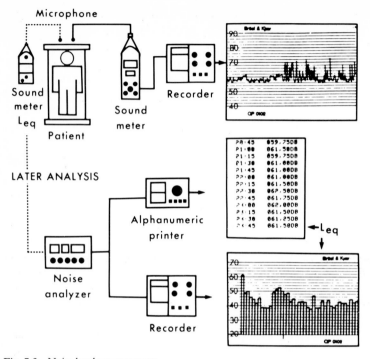

Fig. 7.2   Noise level measurement

using a recorder and digital record form using a printer. For each patient study which was 24 hours long two assistants each observed for a 3 hour period. This was organized in order to cover the 24 hour period for the duration of study in that unit. Observers noted noise sources as they happened. A short interview was aimed at determining their perception of noises and how noise affected them. The settings were intensive care units in three hospitals and ward areas in two of the hospitals.

The data is in the process of analysis (Hilton, in progress). Preliminary analysis seems to indicate that noise varies from ward to ward and hospital to hospital. In some units the noise level was quite high 24 hours a day. In others, the noise level dropped off at night, and in some units it was always fairly quiet.

Why I did what I did

In selecting the design of the study a descriptive approach was used.

An objective method of noise assessment was selected considering the characteristics of noise. A continuous period of study was selected to determine the round-the-clock and round-the-week happenings. Attempts were made to reflect the various types of patient situations in each unit for example, pre-operative, early postoperative, ready for discharge as well as the types of rooms — single, double or ward. In addition the distance from the nursing station was considered. These variables added complexity to the selection of situations and results will be interpreted accordingly.

To locate the microphone to pick up the noises as the patient heard them required various manipulations. The heavy cable, 1½ cm in diameter, was secured up, over and around doorframes and other obstacles to provide for patient/staff safety. For the presence of an assistant to be as inconspicuous as possible, attempts were made to have the machinery based outside the patient room on a cart. In some large units, such as the open-heart recovery room, this was not possible. In one area the observer was situated behind a wall and had to peak out to keep an eye out for what was going on and record it. Attempts were made in all locations to be as least disturbing as possible.

Machinery failure plagued the study and Murphy's Law persisted. This was annoying and trying. It was also time-consuming getting repairs, replacements and making selected alterations to the data-collection plan due to a shortened time-frame. In one situation, during the period the observer was away, the fire alarm sounded. The hospital procedure consists of removing all equipment from the hallways. The equipment was unplugged and removed. When the observer came for the night observation period there was no recorded data.

The equipment was sensitive and required calibration frequently to remain accurate and precise. The degree of need of this was not recognized initially. A set plan for its inclusion was established. Specific checklists had been written to help ensure correct follow-through with all the steps, each one significant to procuring accurate data.

The study will increase understanding of noise in hospitals and how it affects patients. The findings may reveal implications to health care professionals, hospital architects and planners and manufacturers. Some of the units studied will be moving to new quarters lending the possibility of some comparison study in the future.

Diabetic monitoring measures study (Hilton, 1982, in press)

With the advancements in bioengineering, increased use of technology has been incorporated into assessment and therapeutic phases of the nursing process. Both accurate completion of the procedures using the technology, together with appropriate interpretation of the results is important in providing quality nursing care. Capillary blood glucose monitoring (CBGM) is being used more by diabetics as a way of monitoring their disorder as well as by health care personnel to assess and monitor diabetic control. Hospital staff have primarily used urine testing as the method of monitoring control. Are these techniques being carried out accurately? Can we rely on the result which is charted? Do the nurses know the implications of the result? The specific objectives of this descriptive study were:

— to describe the ability of Registerd Nurses (RNs) to perform and interpret diabetic urine testing using the widely used Clinitest 2-drop and Acetest methods;
— to describe the ability of RNs to perform and interpret CBGM using the Chemstrip and Dextrometer methods;
— to describe the ability of RNs to analyze the meaning of capillary blood glucose levels as indicative of hypoglycemia, normoglycemia or hyperglycemia;
— to describe the interrater reliability among RNs in performing diabetic urine testing and CBGM measures;
— to describe difficulties identified in performing and interpreting diabetic urine testing and CBGM methods.

The literature was searched. Few studies described the performance and interpretation of CBGM procedures. Some looked at urine testing or CBGM by patients/staff or studied the correlation of CBGM with laboratory values. There is close correlation when equipment is used according to instructions.

70 RNs from the medical, surgical, intensive care and emergency departments of an acute care hospital in a large metropolitan area were observed performing diabetic urine testing using the Clinitest 2-drop and Acetest methods, and performing CBGM on a volunteer. Nurses who had received inservice on the use of the Dextrometer method were also observed performing this procedure. Subjects were requested to answer a few questions to determine their experience, perceptions, and interpretation of the diabetic monitoring measures.

An assumption that the observed procedure and description of the results in the testing situation would be reflective of what RNs would do in a real patient situation was set.

An observation guide was prepared based on the literature, procedure guidelines and manufacturers' recommendations and tested in a pretest. A panel judged the critical elements in each procedure — those which required accurate completion — since inaccuracy could affect results. A short interview guide was used to gather information on each RN's experience, perceptions and interpretation of diabetic monitoring measures. Test situations using test solutions were provided. Known limits for values from the manufacturer were used for acceptable ranges of results. Results of the Clinitest 2-drop and Acetest methods are presented in Table 7.1. Many of the components were done well. However, considering that each component is critical to the success of the test, one must then be concerned with any deviation from perfect performance. 70% did a perfect test on one specimen but only 30% on the second. 58% did not recognize the occurrence of the pass-through phenomenon. The Acetest procedure was done perfectly by 69%. No statistically significant relationships were found between the number of critical errors commited and the number of times urine testing had been performed by the subject in the past, nor between errors and when the procedure was first learned.

Table 7.1   Critical elements of Clinitest 2-drop method and Acetest method

Critical element	Done (%)	Not done (%)
Clinitest 2-drop method		
1. Clean dry equipment used	96	4
2. Two drops of urine put in Ames test tube with Ames dropper held in vertical position	83	17
3. Dropper rinsed or clean dropper used for water	99	1
4. Ten drops of water put in test tube with Ames dropper held in vertical position	90	10
5. Waited 15 seconds after boiling and then shook test tube gently	86	14
6. Compared color chart and reported glucose level accurately	42	58
Acetest method		
1. Setting one Acetest tablet on a clean dry surface	100	0
2. Dropping one drop of urine on tablet with Ames dropper	99	1
3. Waiting 30 seconds before reading result	72	28
4. Comparing color accurately	90	10

Table 7.2 Critical elements of Chemstrip method

Critical elements	Total group (N = 71) %	Total group (N = 71) Absolute numbers	Inservice group (N = 61) Absolute numbers		No inservice (N = 10) Absolute numbers	
	Done	Not done	Done	Not done	Done	Not done
1. Making sure skin dry after cleansing procedure prior to puncture	37*(26)	63(45)	(20^a+5^b)	(36)	(1)	(9)
2. Applying sufficient blood to cover both patches on Chemstrip	87 (62)	13(9)	(55)	(6)	(7)	(3)
3. Wiping off strip after 60 seconds	80 (57)	20(14)	(49)	(12)	(8)	(2)
4. Matching the squares after 60 seconds (55–65 accepted)	96 (68)	4(3)	(52)	(9)	(8)	(2)
5. Reported value correctly by matching either the correct score or correct range	63 (45)	37(26)	(42)	(19)	(3)	(7)

[a] allowed 15 seconds for alcohol to dry
[b] used other cleansing methods
* 37% = 7% a + 30% b

For the Chemstrip method the results are indicated in Table 7.2. The data was analyzed together and separately for those who had or had not the formal inservice program on the Chemstrip procedure. For the total group 26.8% made no critical errors, 33.8% made one and the remainder made more than one.

Only 39 subjects performed the Dextrometer testing measures since all nurses had not attended the inservice program on the use of this method; 11 critical elements were identified; 16 made no critical errors and 10 made one.

Charting for all procedures was not consistently well done. It was concluded that since RNs from only one acute care hospital were tested and the investigation was descriptive in nature, no generalizations to other areas could be made. However, it could be concluded that the RNs in that hospital had difficulty performing the monitoring measures accurately. The findings had several implications for nursing education and practice. Since the nurse has a role in teaching the procedure and its interpretation to the patient, she too must be aware of the difficulties and be able to perform and interpret these methods well. The meaning, limitations, and idiosyncrasies of each test should be taught in addition to procedural specifics so that nurses can make sound judgements for patient care.

Why I did what I did

Although observation of a patient situation might have provided more information, this was not feasible because the data collection period would have taken four times as long. Instead it seemed important to set up a simulated situation with built-in controls; this also provided an objective measure of accuracy because of known levels for both urine and blood solutions. On research assistant did all of the data collection. This provided consistency and eliminated interrater reliability problems. The use of a specific checklist helped maintain consistency. There is the possibility that her observation skill altered during the period and that this influenced results. Timing of the parts of the procedure which require accurate timing was done with a stopwatch held inconspicuously. Several subjects later commented they had not realized a stopwatch was used. The test situation was located in a patient room on an unused ward therefore making it similar to an actual patient situation. Organizing times to test subjects became a major endeavour. Appointments were made but often had to be cancelled depending on the ward needs.

Analysis of the data was completed using a computer. This was the first time the researcher had used a computer for this purpose. It proved a challenge. Computers, like most electronic devices, must have their commands given in a precise way. For example, if a space must go between specific parts of the command, one inaccuracy will result in the production of either nothing or garbage. The necessity of making a duplicate data file was learned painfully when, for some unknown reason to the computing center, the data was accidentally erased; luckily, it was retrieved through a complicated process.

SUMMARY AND CONCLUSION

Electronic instrumentation may be the most appropriate data-collection method to answer the question. This chapter has reviewed electronic instrumentation and some of the knowledge, concerns, headaches and joys of using it have been discussed. In illustrating these concepts the author's studies of sleep, noise and performance of diabetic monitoring measures were described. Rationale was provided for methodology selection and aspects of implementing the data-collection and analysis phases were shared. Ideally, through this discussion, the reader has increased knowledge about indicators, about data-collecting tools which incorporate mechanical instrumentation, about the ways these are used in evaluating aspects of patient care, and about the use of these tools in giving care.

Research findings have contributed significantly toward advancing knowledge, technology and nursing practice. To use nursing research to improve care, nurses should strive to discover and refine the nursing practices which are most effective in meeting the needs of people requiring nursing services. Research should influence practice otherwise it gathers dust and is of no use to anyone. Problems which are pertinent and relevant to practice ought to be ones chosen for study. Research should not be a thing unto itself; research should not exist for research's sake. Quality of care can be improved when we all get together: let us bridge the gap!

REFERENCES

Abbey J C 1975 Physical indicators in clinical nursing research or why do I need to know what? In: Development and use of indicators in nursing research. Proceedings of 1975 National Conference on Nursing Research, Edmonton, p 31–43

Fox D J 1982 Fundamentals of research in nursing, 4th edn. Appleton-Century-Crofts, Norwalk, Connecticut

Geddes L A, Baker L E 1969 Principles of applied biomedical instrumentation.
Wiley, New York

Hilton B A 1976 Quantity and quality of patients' sleep and sleep-disturbing factors
in a respiratory intensive care unit. Journal of Advanced Nursing 1: 453–468

Hilton B A 1982 Diabetic monitoring measures — does practice make perfect?
Canadian Nurse 78(5): 26–32

Hilton B A 1982 Does diabetic control really make any difference? Canadian Nurse
78(9): 49–52

Hilton B A Nurses' performance and interpretation of urine testing and capillary
blood glucose monitoring measures. Journal of Advanced Nursing (in press)

Hilton B A Noise levels in hospitals. (in progress)

Jacobs M R 1978 Sources of measurement error in noninvasive electronic
instrumentation. Nursing Clinics of North America 13(4): 573–587

Luce G G 1971 Body time. Pantheon, New York

Rechtschaffen A, Kales A, 1968 A manual of standardized terminology, techniques
and scoring system for sleep stages of human subjects. U.S. Department of Health,
Education and Welfare, Maryland

Stinson S 1973 Staff nurse involvement in research — myth or reality? Canadian
Nurse 69(6): 28–32

Treece E W, Treece J W 1977 Elements of research in nursing, 2nd edn. Mosby, St
Louis

Treguna J R, 1977 Overview of the RNABC quality assurance program. Registered
Nurses' Association of British Columbia, Vancouver

Ventura M R, Hinshaw A S, Atwood J R 1981 Instrumentation: the next step.
Nursing Research 30: 257

Walker J G 1978 Noise in hospitals. In: May D N (ed) Handbook of noise
assessment. Van Nostrand Reinhold, New York, ch 7, p 185–194

The use of computers for improvement and measurement of nursing care

The delivery of nursing care in the most cost-effective way within a variety of health care systems, presents a challenge to the nursing profession which has up to the present been difficult to meet. The reason for this is twofold; the first lies in the difficulty in capturing the vast amount of data generated by nursing procedures and converting it to an acceptable index of nursing workload. The second is the absence of a satisfactory methodology for assessing the quality of nursing care. The second problem cannot be tackled without first solving the first. The proposition put forward in this chapter is that the solution lies in the application of computer technology to the delivery of nursing care associated with the automatic data capture which is a by-product of the technology. It is proposed therefore to describe here a real-time computerised nursing system primarily designed to facilitate the optimum delivery of nursing care to patients and the way in which the resulting automatic data capture has been used to arrive at indices of nursing workload which may be exploited to evaluate the quality of nursing care.

COMPUTERISED REAL-TIME NURSING SYSTEM

A real-time nursing system has been operating on medical and surgical wards at Ninewells Hospital, Dundee, for the last 6 years. It comprises a patient administration system and a nursing system, used by the nurses. Patient data is entered by the nurse in charge of the ward using a visual display unit (VDU) linked to a Modular I computer.

The system relies on selection by the charge nurse of items from a standard precoded list of nursing orders, specimen collections and tests which allows a daily care plan for each patient to be generated on a VDU and printed on A4 size paper. The care plan is placed at the foot of the patient's bed to give an immediate visual notification of required care and to allow convenient recording of the care given.

An important benefit is the opportunity to provide the documentation of care on a patient-orientated basis.

The system's potential to deliver effective nursing care to patients is illustrated by the way it can be organised and documented, and the records utilised to give basic and technical nursing care, to provide continuity (while in hospital and on discharge), to facilitate educational programmes and to assist administration.

NURSING CARE PLAN

Using the computer system, the charge nurse identifies the patient's problems and the appropriate nursing solutions. She then obtains access to the system through a VDU (Fig. 8.1) and generates a nursing care plan. An example of such a care plan is shown in Figure 8.2 and concerns a 72-year-old female patient who had sustained a hemiparesis 2 weeks previously. She was hypertensive and had an anaemia, suspected to be either pernicious anaemia or due to chronic gastrointestinal blood loss. Her nursing care was planned in the light of her problems. For example, special attention was paid to oral hygiene, pressure areas, rehabilitation and blood pressure recording. The nursing procedures listed on the printed care plan are linked with boxes where the times of the procedures and the nurse who performed them can be entered. On completion of the procedures the nurse in charge of the ward therefore has available to her a record of implementation of the plan and can identify the nurses who have

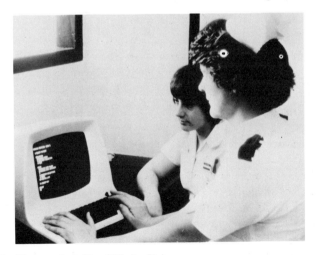

Fig. 8.1   Nurses using a Visual Display Unit

carried it out.

The system aids the performance of various investigative procedures by the nursing staff. For example, in the case of this anaemic
patient, a stool test for faecal occult blood and a Schilling test to
measure vitamin B_{12} absorption have been ordered. For the latter
test, the computer generates a cascade of timed procedures to assist
the nurse as shown in Figure 8.2.

The computer can be used in other situations, such as preparation
for radiologic or endoscopic investigation. In addition, lists of patients who have to be fasted at different times can be printed-out
from the computer memory.

The advantages to nurses conferred by the system can be appreciated in the light of the complexity of modern diagnostic methods.
Furthermore, the well-organised record improves communication
between nurse and patient, and reduces the possible timing errors
that can unnecessarily extend a patient's hospital stay. The patient-
orientated nursing record facilitates the delivery of nursing care by
the same nurse or nursing team and has obvious benefits in improving nurse-patient relationships.

NURSING CARE SUMMARY

All nursing care given to a patient is stored chronologically and

```
WARD 08                   PATIENT CARE PLAN            12:26  19/01/81

GILBERT    SUSAN          1711630078        DAY 3

                          START                   INITIALS              ! NIGHT STAFF !MANUAL BACKUP SYSTEM
                                          AM              PM             ! SIGNATURE  !
                                      8    10   12    2    4    6    8   !            !

NURSING ORDERS
  BED BATH                19/01/81
  MOUTH WASH              19/01/81
  PRESSURE AREAS 2 HRLY   19/01/81
  TOILET - ASSISTANCE     19/01/81
  HAIR CARE               19/01/81
  EYE CARE                19/01/81
  WEIGH MON WED SAT       19/01/81
  4-HRLY T, P AND BP      19/01/81
  PHYSIOTHERAPY           19/01/81
  LIGHT DIET              19/01/81
  WALK WITH ASSISTANCE    19/01/81
  DAILY ABDO. MEAS'M'T    19/01/81
  OCCUPATIONAL THERAPY    19/01/81
  4-HRLY LYING/STAND BP   19/01/81

SPECIMEN COLLECTIONS
  URINE - M.S.S.          19/01/81 EM
  URINE - M.S.S.          20/01/81 EM
  URINE - M.S.S.          21/01/81 EM
  STOOL TEST FOR F.O.B.   19/01/81 DD

TESTS
  SCHILLING - PLAIN       19/01/81 EM
    8 AM - PATIENT VOIDS BLADDER
           COMMENCE 24 HOUR URINE COLLECTION
           ADMINISTER ONE CAPSULE TO PATIENT
    10 AM - ADMINISTER INJECTION HYDROCOBALAMINE 1000 MICROGRAMS
    8 AM TO 12 MIDDAY - PATIENT CAN DRINK WATER IF DESIRED
    12 MIDDAY - PATIENT MAY TAKE SOLID FOOD
```

Fig. 8.2   Nursing care plan

```
WARD 06              PATIENT CARE SUMMARY        10:53  02/03/84

BLACK      MARY       2310436723  AGE 40

START STOP                                    START STOP

-----------BASIC NURSING CARE----------------------   ------------HORILITY----------------------------
28/02 29/02 ORAL HYGIENE - 2 HOURLY            28/02 01/03 BED REST
28/02 01/03 PRESSURE AREA CARE - 2 HOURLY      01/03 02/03 UP FOR 30 MINS TWICE DAILY
29/02 01/03 ORAL HYGIENE - 4 HOURLY            02/03       UP FOR 1 HOUR TWICE DAILY
28/02       BED BATH
28/02       HAIR CARE                          ------------SPECIAL NURSING CARE----------------
01/03       MOUTH WASH                         28/02 29/02 CARDIAC MONITOR
01/03       PRESSURE AREA CARE - 4 HOURLY      28/02 29/02 INTRAVENOUS FLUIDS AS PRESCRIBED
                                               28/02 29/02 OXYGEN AS PRESCRIBED
-----------TOILET------------------------------   28/02       FLUID CHART
28/02 02/03 COMMODE                            02/03       BED CAGE IN SITU
02/03       TOILET - ASSISTANCE
                                               ------------FEEDING------------------------------
-----------TP AND BP----------------------------   28/02 29/02 FEED WITH ASSISTANCE
28/02 29/02 HOURLY TEMPERATURE/PULSE/BLOOD PRESSURE  29/02       FEED BY SELF
29/02 01/03 2-HRLY TEMPERATURE/PULSE/BLOOD PRESSURE
01/03       4-HRLY TEMPERATURE/PULSE/BLOOD PRESSURE  ------------DIET---------------------------------
                                               28/02 01/03 LIGHT DIET
-----------WEIGHT-------------------------------   01/03       ORDINARY DIET
29/02 02/03 WEIGH DAILY
02/03       WEIGH WEEKLY


                               START       STOP
SPECIMEN COLLECTIONS                                DATE--WEIGHT----DATE--WEIGHT----DATE--WEIGHT----
                URINE ROUTINE TEST  28/02/84 DD   28/02/84   29/02 52.4      01/03 52.3    02/03 52.2
                BLOOD CULTURES      28/02/84 DM   28/02/84
                URINE - M.S.S.      28/02/84 DD   28/02/84  DATE--S.G.--PH-SUGAR-PROTEIN--------KETONES-BLOOD--BILE-
                URINE ROUTINE TEST  02/03/84 DD   02/03/84   28/02 1015  6  NEG.  30MG.(+)       NEG.   NEG.  NEG.
TESTS                                                        02/03 1016  7  NEG.  NEG.           NEG.   NEG.  NEG.
                CHEST X-RAY         28/02/84 DD   28/02/84
                ELECTROCARDIOGRAM   28/02/84 DD   28/02/84
                ECHO-CARDIOGRAM     06/03/84 14:00


ALL NURSING CARE COMPLETED:.......................(SIGNATURE)
```

Fig. 8.3 Nursing care summary

recalled on a VDU or printed-out when required. The summary provides a continuous record of all nursing care and in addition includes nursing observations, e.g., mental state, auditory and visual defects, profiles of body weight and urine analysis (Fig. 8.3).

In the last decade, the organisation of nursing care has changed radically, both in an administrative and operational sense. A major organisational problem has resulted from a reduction in nursing hours and the requirements of new nursing curricula. The revised nurse scheduling caused by these developments has produced difficulties in the provision of continuous nursing care. There has been an increase in 'communication interfaces' — a nurse who takes over the care of a patient must be provided with information about the nursing care planned. The legible and well-organised plan provided by the computer-generated summary minimizes such difficulties and complements verbal reporting.

Another problem militating against continuity of care is speciialisation, which often leads to frequent transfers of patients between wards and hospitals. The breaks in continuity produced by such patient movements are eliminated in wards using the computer generated summaries which can be transmitted instantaneously. If transfers of nursing data are required to wards, hospitals, or community nursing services not using the computer, then the nursing summary is printed-out.

Educational applications

In recent years, the role of the nurse tutor in nurse education has increased. However, the tutor has special difficulties when teaching on the ward since clinical teaching has to be opportunistic and educational and service objectives are frequently in conflict. These problems can be overcome by a co-operative ward sister, but in wards where the service pressures are great, adequate communication may be difficult.

The computerised current nursing care plan and summary allows the tutor to teach effectively without having to make unreasonable demands on the time of the charge nurse. Another important contribution to the work of the nurse tutor is the patient-orientated record. The value of these records to the in-service training of nursing staff cannot be underestimated. The data-recall by the VDU during reporting and care-planning sessions can also have both an educational and service role.

Administrative applications

The charge nurse can plan and implement nursing care in a systematic way using a computer and since the plan provides a good record making effective implementation more likely, she can readily ascertain the current nursing care on returning from an off-duty period.

The reduction in administrative workload produced by the system can give charge nurses more time for direct involvement in patient care, allowing them to make more adequate assessments of patient's psychological and social problems — so important in delivering effective care.

Nurse scheduling application

The problem of devising a practical method of allocating nursing staff in hospitals to various wards or services with the objective of producing an optimum match of nurses with appropriate experience to the amount and nature of the workload has proved, over the years, to be of great complexity. One of the factors contributing to the difficulty of solving the problem has been the lack of a satisfactory index of a nursing workload. The most successful attempts to arrive at estimates of workload, the Aberdeen Formula (Scottish Home & Health Department, 1969) and the 'Barr Index' (Barr, 1967) have a number of defects, the most important of which are their dependence on subjective estimates of patient dependency. Furthermore all manual systems previously used in this field encounter major logistic problems due to the large amount of data generated and the consequent difficulties of data capture, analysis and interpretation. The availability of the real-time computerised nursing system in medical wards of a large teaching hospital described above, and in more detail by Henney et al (1977), provided firstly the opportunity of developing a method of deriving an index of nursing workload and secondly a flexible data processing facility. This could make a contribution to the solution of these problems and hopefully point the way to the achievement of more satisfactory nurse allocation procedures.

The problem of obtaining a measurement of nursing workload was tackled by devising a series of indices of nursing workload using the data captured by the computer system. These evolved from an original index based on the opinion of one sister, regarding the load generated by the individual nursing procedures and from indices reflecting the corresponding opinions of a large number of sisters from several Scottish hospitals. Finally it took the form of an index derived from the conversion of the rating system given by the sisters to timings based on the results of previous work studies of individual nursing procedures. These indices are shown in Table 8.1 in the order they were developed and a more detailed account of their development follows.

NINEWELLS INDEX 1

The automatic data capture facilitated by the computer was exploited by developing an index of nursing workload (Ninewells Index 1) for general medical wards based on the assignment of a

Table 8.1   Developed indices of nursing workload

Index	Ninewells I	Delphi I	Delphi II	Ninewells II
Type of rating	0–5 point scale	0–5 point scale	Interval scale ratings	Timings
Basis for deriving ratings	1 nurse from 1 hospital	112 nurses from 33 hospitals	Transformation of Delphi I ratings	39 nursing procedures 3 workstudies

score to each nursing order on a 5-point scale. This when summed for all patients on a ward would give an index of the nursing workload for that ward. The number of points assigned to each nursing procedure was based on the opinion of an experienced nurse. The number of nursing orders in the system is over 200 including tests and specimen collections. Moreover the nursing procedures were divided into the basic and technical ones, the latter being defined as those not normally carried out by untrained staff. This method of assessing nursing workload has been shown to be very sensitive to workload changes by Henney & Bosworth (1980). Using the Ninewells Index 1 the relationship between basic and technical care, which is regarded by some workers as constant (Scottish Home & Health Department, 1969), was investigated, as was the relationship between the allocation of nurses and the workload in the six medical wards using the computer system. The results have been reported by Henney et al (1979) and showed an absence of correlation between them. This conclusion is dependent upon the validity of the Ninewells Index 1 as an estimate of nursing workload and in order to test this assumption a survey was mounted to establish agreed ratings for standard nursing procedures by a large number of experienced charge nurses in the medical wards of teaching and district general hospitals in Scotland.

DELPHI INDEX I

The method adopted for validating the Ninewells Index I was the Delphi technique developed by the Rand Corporation and designed to obtain the most reliable consensus of a group of experts (Dalkey, 1969). The Delphi technique consists of a series of questionnaires which initially contain a number of primary questions directed at the experts and followed by feedback of the results of all experts to each individual who then produces a modified set of responses to the same questions. This mode of controlled interaction among the experts is a deliberate attempt to avoid the disadvantages associated with more conventional uses of experts such as in round table discussions or direct confrontation of opposing views. Typically, the answer to the primary question is a numerical quantity (in this study the rating for each nursing procedure). It is expected that the individual expert's estimates will tend to converge as the experiment continues even if the estimates expressed initially are widely divergent.

The Delphi study was conducted with a panel, consisting of 115 medical ward sisters from 33 Scottish hospitals. The aim of the study, as previously mentioned, was to arrive at a consensus regarding the workload ratings of the nursing procedures on the 0–5 scale. With the assistance of the Chief Area Nursing Officers, the ward sisters who were to participate in the study were approached, and their co-operation was achieved.

The first round of the study involved the production of letters describing the aim of the study and asking each sister to rank certain items of nursing care on a 5-point scale. It was emphasised that these ratings should not be based on individual patients, or the condition of the patients, but on a comparative estimate of the average load produced by the individual procedures. It was also explained that the dependency status of the patient was taken into account by weightings for mental state, auditory and visual defects and incontinence. Due to the large number of procedures in the computer system, the sisters were randomly divided into groups, each group having a sub-set of about 85 nursing tasks to rate. In order to avoid biased ratings, these sub-sets of nursing procedures overlapped to the extent of 43 procedures, ensuring thus that each task would be rated by two groups. However a full list of the nursing procedures in the computer was also sent to all ward sisters so that they had an overall picture of the procedures. Of the 115 questionnaires sent out, 112 were returned completed.

The second part of the study involved feedback to the participants of the first round and was initiated by sending a second letter to each sister describing the principles of the Delphi method and explaining that they could now change their original ratings if they wished to do so in view of the ratings of their colleagues. Together with this letter a questionnaire was sent for recording any revision of ratings. Of the 112 questionnaires, 102 were returned completed.

The results of the first round of the survey showed that although the conditions which prevail in each hospital and the way each nurse perceived the various nursing procedures may have affected their ratings, a reasonable degree of agreement between the participating sisters existed. Not surprisingly, this was particularly true for basic as opposed to technical nursing procedures.

Table 8.2 shows the ratings assigned to a representative sample of nursing procedures in the first round of the Delphi study. Since the participants were asked not to rate procedures which were not carried out in their wards, the number of gradingss for each procedure varied. In many cases a substantial majority agreed on a particular rating as in 'mouth wash', 'bath with assistance', and to a lesser extent in 'Diabetic urinalysis'. However, even when such a

Table 8.2   Rating form used after the first round of the Delphi study

Number of your colleagues choosing grade 0–5 0 1 2 3 4 5						Nursing orders	Your first grading	Your second grading
0	2	19	12	8	4	Toilet-assistance	...	...
22	16	2	1	0	1	Toilet-independent	...	...
3	5	7	6	1	0	uridome	...	...
2	21	4	2	1	0	urinal	...	...

majority did not occur the ratings tended to be found in procedures such as 'catheter intermittent' and 'arteriogram'. The latter represents a category of nursing tasks, mainly technical, where there was widespread disagreement. Such tasks, however, occurred with low frequency in the general medical wards included in the study, and this was reflected by the small number of participants who rated them.

During the second round many sisters changed some of their ratings in the light of their colleagues' opinions and a much closer agreement was achieved. An example of such changes is given in Table 8.3.

From this survey, new ratings were derived for each nursing procedure based on the mode (majority rating) after ascertaining that other indices based on mean and median values showed no statistically significant difference. The new index so derived was named the Delphi Index 1.

DELPHI INDEX II

The data collected by the Delphi survey although essential for validating the approach adopted for estimating the nursing workload, was based (as the original ratings) on an ordinal 0–5 point scale. Therefore it could be argued that the resulting index, even if it is

Table 8.3   First and second round ratings of nursing procedures

Nursing procedure	0	1	2	3	4	5
Bath with assistance	0(0)	1(1)	10(8)	18(21)	9(9)	8(7)
Mouth wash	2(1)	34(36)	7(7)	2(1)	1(1)	0(0)
Catheter intermittent	1(1)	5(4)	13(14)	16(16)	7(7)	2(2)
Diabetic urinalysis	0(0)	9(6)	18(22)	8(7)	10(10)	2(2)
Arteriogram	2(2)	5(5)	2(2)	6(6)	6(6)	3(3)

N.B. Numbers in brackets indicate second round ratings.

valid, cannot support any studies for establishing methods to achieve a better utilisation of the existing nursing resources. This is because an ordinal scale can support comparisons of the type 'A is greater than or equal to or less than B' but cannot quantify the difference between these two quantities. This problem can be overcome by processing further data gathered by the Delphi survey so that the ordinal ratings can be transformed to ratings based on an interval scale, which can quantify differences between two items. The method used for that transformation is based on the 'Law of Categorical Judgement' developed by Thurstone (Torgerson, 1958). This method assumes that although people do not always react in the same way and therefore neither rate an object with the same value on a given scale, nor perceive the categories into which that scale is divided in a similar manner, their responses form a statistical (normal) whose mean value is the required rating of the object on an interval scale. An additional advantage of having ratings on an interval scale is that a comparison with timings of procedures can be made so that more conclusive evidence regarding their validity can be obtained. Also, if a strong relationship between ratings and timings is proved to exist, it can be used to transform the former into the latter and thus obtain a potentially more useful index of nursing workload, based on the time required to carry out the workload.

The statistical tests (Kolmogorov-Smirnov) which were carried out, showed that the ratings resulting from the application of the 'Law of Categorical Judgement' represented the opinion of the participating sisters with acceptable accuracy and only eight tasks (out of 212) did not follow this pattern. The failure of the model to fit the data in those eight cases was due to a few ratings (amounting to 5% of the total number of responses) for each of the eight tasks. These could be omitted without distorting significantly the data and statistically acceptable results obtained for those cases too.

NINEWELLS INDEX II

The investigation regarding the validity of the ratings of the Delphi Index II (which, as mentioned before, can be carried out by comparing them with the corresponding timings) was based on a sample of 39 representative nursing procedures selected to span a wide range of workload. The timings for those procedures were obtained from work studies carried out in England and Scotland in medical/geriatric wards (Grant, 1977; Rhys Hearn, 1978) which in that respect were similar to the medical wards in Ninewells Hospital. Figure 8.4

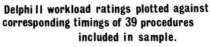

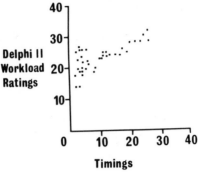

Fig. 8.4   Delphi II Workload ratings plotted against corresponding timings of 39 procedures included in sample

shows the Delphi II workload ratings plotted against the corresponding timings of the 39 procedures in the sample.

It can be seen that with the exception of seven tasks the ratings bear a strong relationship with the timings. The correlation co-efficient is as high as 0.75 and rises to 0.91 if the above seven tasks are omitted. The nursing orders which do not fit this pattern have high Delphi ratings and low timings. It is believed that this disagreement is due to differences in the interpretation of these tasks and the inclusion in the assessment of their workload by the sisters of other related procedures which were considered separately by the work studies.

The existence of such a definite correlation between ratings and timings allows the transformation of the former into the latter (with an average absolute error of 3.2 minutes and standard deviation of 2 minutes) and thus, after some minor adjustments of the generally realistic results to reflect local conditions, an index of nursing workload in terms of time was obtained and named Ninewells Index II. Subsequently, in order to take into account the clerical workload of the wards and other activities such as drug rounds and doctors' rounds, an estimated amount of time was added, so that the index could give an overall picture of the load imposed on each ward. The time required for these activities was estimated by analysing further data which was collected during a study, carried out in the medical wards in Ninewells Hospital jointly by the Health Services Operational Research Unit of the University of Strathclyde and the Nursing Research Unit of the University of Edinburgh (Collings, 1977).

Comparison between indices

A first comparison of the Ninewells I and the Delphi I indices showed that the consensus of the participating sisters about the comparative workload involved in each nursing procedure was close to the one expressed by the Ninewells Index. As can be seen from Table 8.4, the workload ratings of a substantial proportion of nursing procedures were identical for both indices and that when differences did occur, they were usually of the magnitude ± 1. A variety of statistical tests showed that for the bulk of the nursing procedures, the basic ones, no consistent bias between the two indices existed. In the area of technical care, the Ninewells Index I tended to overestimate the specimen collection and to a lesser extent the tests, which, however, do not contribute markedly to the total workload.

The comparison of the two indices takes no account of the frequency of each nursing procedure which is a crucial factor in determining the overall accuracy of the estimated workload. Henney & Bosworth (1980) suggested that with an error of ± 1 in the workload ratings of the nursing tasks, the error of the estimated workload would be less than 5%. In order to establish the likely magnitude of the differences between the overall workload estimated, calculated by the Delphi I and Ninewells I indices, the total workload of the six wards in the study was observed on 36 randomly selected days, six for each ward and computed using these two indices. The results showed that the differences were small and on average 3.77%. The Ninewells Index II was subsequently used in a further study of similar design to that described above, to investigate the validity of the Ninewells Index I which as shown before can be regarded equivalent to the Delphi Index I. The results of the comparison of the daily workload estimated of the two indices cover a period of over 2 months and Table 8.5 shows the correlation co-efficient between the two measures of the basic, technical and total workload of the six wards, and demonstrates satisfactory equivalence. These results which show exceptionally high correlations between the two

Table 8.4   Delphi I vs Ninewells I: differences between corresponding ratings

Magnitude of difference	+3	+2	+1	0	−1	−2	−3	Total
Basic nursing tasks	1	10	21	37	19	8	1	97
Technical	3	4	7	45	36	16	5	116

Table 8.5   Ninewells II vs Ninewells I: correlation co-efficients

Ward	1	2	3	4	5	6
Basic workload	0.90	0.92	0.94	0.90	0.73	0.98
Technical workload	0.90	0.92	0.78	0.86	0.83	0.82
Total workload	0.94	0.93	0.93	0.81	0.81	0.98

indices for all types of workload suggest that the Ninewells Index I is a valid index which reflects accurately the workload of the wards and their comparative needs in nursing staff. Reversing this argument, it may also be suggested that the Ninewells Index II is valid since it reflects the opinion of the sister who developed the Ninewells Index I, which in turn, coincides to a great extent with the opinion of the 112 sisters who participated in the Delphi survey.

The Delphi study, as stated before, had the aim of evaluating the validity of the ratings of the Ninewells Index I, which were based on the opinion of only one ward sister, by comparing them with other ratings derived from the opinion of more than 100 other ward sisters from 33 hospitals and working under different conditions. It is of interest that among the participants, fairly good agreement of their workload ratings was achieved at the first round and a relatively better one during the second round which showed that the principles on which the Delphi method is based operated satisfactorily.

The results of the study have shown quite clearly that the original Ninewells Index I provides results, in estimating nursing workload, comparable with those obtained by the Delphi method (Delphi Index I). Differences between the ratings of the two indices certainly did exist but significantly they were mainly in the area of more specialised nursing care of tests and specimen collections. The reason for this difference may be attributed to the fact that these tasks are more affected by the local conditions and also by variations in the degree of participation and involvement of nursing staff in carrying them out in different hospitals. These sources of difference in workload ratings became apparent from enquiries of some of the participants during the two rounds of the study. However, as these procedures are not carried out very frequently, and as there was a high degree of agreement for the more common ones the observed discrepancies in particular ratings did not affect significantly the accuracy of the total workload.

The interval scale ratings, which were established using the data collected during the 'Delphi' study, provided an index (Delphi Index II) which can overcome the mathematical problems presented by the previous ordinal scales. In addition these ratings proved to represent the general opinion of the sisters who participated in the Delphi survey with acceptable accuracy and thus it may be argued that they reflect the relative workloads. Further evidence regarding the validity of this approach was obtained by comparing the ratings based on Delphi Index II and the timings of a representative sample of nursing procedures, and demonstrating that a strong relationship exists between those two measures of the nursing workload.

The strong correlation between ratings and the timings facilitated also the transformation of the Delphi II ratings to timings and the establishment of a Ninewells Index II based on a time and thus potentially more useful than the other indices.

DISCUSSION

The real-time computerised nursing system described in this chapter has contributed to improvement in the nursing care of patients and has also made possible a measurement of nursing care (the workload) which could facilitate improvement of that care by better matching of nursing resources to the needs of patients.

The ultimate test of any nursing system, in an administrative sense, is its acceptability by nurses whose motivation lies in looking after patients to the best of their ability. The computer system has been operational for more than 6 years and due to nursing demand has been introduced to all medical wards in our hospital and has now been implemented in surgery. This stepwise implementation of computerisation in nursing has allowed the system to evolve in terms of the real-life requirements of patients in a variety of different medical environments. This approach is likely to improve the quality of nursing care although it must be accepted that the methodology of proving this hypothesis remains to be developed. However, the identification of nursing problems when linked to the accurate and unambiguous (and legible) nursing care plan generated by the computer system will make practical evaluation of the efficacy of nursing care using standard clinical trial methodology or case-control design. Nursing research on the effectiveness of nursing care on a large enough scale will only be feasible by computerisation. Such research could transform nursing practice and so benefit patients in the same way in which the clinical trials and case-control studies of drugs have done.

The measurement of nursing workload and the matching of workload to nursing resources is an area of research endeavour which has been relatively unsuccessful due to the inability of manual methods to cope with the vast amount of data to be handled quite apart from identifying the units of measurement to be used. The potentiality of the computer to facilitate research in this field is hopefully illustrated by the results reported in this chapter and if demonstrated to be successful in optimising the matching of nursing care to patient needs will make a major contribution to improving the quality of nursing care. This is the next objective of our research endeavours and its achievement will confirm the view that in the computer, nurses have a more powerful tool than any previously available to them in maximising the effectiveness of nursing care.

REFERENCES

Barr A 1967 Measurement of nursing care. Publication No. 9, Operational Research Unit, Oxford Regional Hospital Board

Bosworth R, Henney C R, Crooks J 1980 A computer based system for the automatic production of nursing workload data. Nursing Times 76(28): 1212–1217

Collings T 1977 An evaluating of the ward computer project at Ninewells Hospital. HSORU, University of Strathclyde

Dalkey N C 1969 The Delphi method: an experimental study of group opinion. RM-5888-PR, Rand Corporation, Santa Monica

Grant N 1977 A method for calculating nursing workload based on individualised nursing care. University of Edinburgh PhD Thesis

Henney C R, Bosworth R, Brown N, Crooks J 1977 Can a computer improve communications in the wards area? Medinfo 77: 953–956

Henney C R, Bosworth R, Chrissafis I, Crooks J 1979 Nurse allocation by computer. Proceedings of The International Conference In Medical Computing, Berlin

Rhys Hearn C 1978 Private communication, Department of Social Medicine, University of Birmingham

Scottish Home And Health Department 1969 Nursing Workload Studies. Scottish Health Service Studies No. 9, SHHD, Edinburgh

Torgerson W C 1958 Theory and methods of scaling. Wiley, New York

PART FOUR

Bibliography

1. General discussions

In this section readers will find discussions of basic ideas, issues and an indication of the scope of the writings on quality of nursing care. Also there are several collections of items on quality assurance or on development of criteria (Nursing Clinics of North America, June 1984, a symposium edited by Zimmer; Downs & Newman, 1977; National Conference on Nursing Research, 1975). The latter contains keynote addresses by Lisbeth Hockey, Jack Hayward and Jane Abbey on social, psychological and physical indicators respectively.

Bailet H, Lewis J, Hochheiser L, Bush N 1975 Assessing the quality of care. Nursing Outlook 23(3): 153–159

Bloch D 1975 Evaluation of nursing care in terms of process and outcome: issues in research and quality assurance. Nursing Research 24: 256–263

Bloch D 1977 Criteria, standards, norms: crucial terms in quality assurance. Journal of Nursing Administration Sept: 20–30

Bloch D 1980 Interrelated issues in evaluation and evaluation research: a researcher's perspective. Nursing Research 29: 69–73

Crow R 1981 Research and standards of nursing care: what is the relationship? Journal of Advanced Nursing 6: 491–496

Donabedian A 1976 Some basic issues in evaluating the quality of health care. In: Issues in evaluation research. American Nurses Association: 3–28

Downs F, Newman M (eds) 1977 A source book of nursing research, 2nd edn. F A Davis, Philadelphia, Part I

Downs F, Zimmer M 1980 Relationship of findings of clinical research and development of criteria: a researcher's perspective; a nursing service administrator's perspective. Nursing Research 29: 94–99

EURO 1979 Reports and Studies #4 (1977–1979) Evaluation of inpatient nursing practice. Report on a working group. Regensburg 18–21 October 1977. Regional Office for Europe World Health Organization, Copenhagen

Hartman M 1976 An historical perspective on quality assurance. In: Pathways to quality care. National League for Nursing: 1–5

Hegyvary S 1979 Nursing process: the basis for evaluating the quality of nursing care. International Nursing Review 4(26): 113–116

Hunt J 1981 Indicators for nursing practice: the use of research findings. Journal of Advanced Nursing 6: 189–194

Inman U 1975 Towards a theory of nursing care. Royal College of Nursing, London, section 2: 47–111

Janforum: The nursing process and standards of care 1981. Journal of Advanced Nursing 6: 503–514

Kurowski B, Breed S 1981 A synthesis of research on client needs assessment and quality assurance programs in long-term care. Western Interstate Commission for Higher Education. Boulder, Colorado, USA

Lang N 1974 Are professional nurses ready for a quality assurance program? Journal of the New York State Nurses' Association 5(4): 24–32

Lindeman C 1976 Measuring quality of nursing care: review of current research projects. Journal of Nursing Administration 6: 16–19

Lindeman C, Brush R, Hanson R, Hinshaw A, Holloway J, Krueger J, Loveridge C, Lum J, Nelson A 1978 An empirical approach to defining quality of nursing care. In: Communicating Nursing Research 2: Western Interstate Commission for Higher Education, Boulder, Colorado, USA: 32–34.

National Conference on Nursing Research: Proceedings 1975 Development and use of indicators in nursing research. University of Alberta, Edmonton, Alberta, Canada

Nicholls M E, Wessells V G (eds) 1977 Nursing standards and nursing process. Contemporary Publishing Incorporated, Wakefield, Massachusetts, USA

Padilla G, Grant M 1982 Quality assurance programs for nursing. Journal of Advanced Nursing 7: 135–145

Ramey I 1973 Setting nursing standards and evaluating care. Journal of Nursing Administration 3(May–June): 27–35

Roper N 1976 A model for nursing and nursology. Journal of Advanced Nursing 1: 219–227

Stevens B J 1972 Analysis of trends in nursing care management. Journal of Nursing Administration 2(November–December): 1–6

Zimmer M (ed) 1974 Symposium on quality assurance. Nursing Clinics of North America 9(June): 303–379

2. Nurse performance

Readings in this section are aimed at improving the performance of the individual nurse who is giving direct patient care. Measurement which is used for assessment or for showing where improvement is needed is described using two main approaches: measurement by an administrative process using an established standard (Sheridan, Fairchild, Kaas; Schwirian; Haar & Hicks) and peer review procedures. Peer review was a major focus of interest in the late 1970s. Tools developed for management or peer review often can be used for self appraisal. In the next section on Clinical Studies an article by Harrison presents self evaluation tools while one by Heidt deals with the individual nurse's therapeutic use of self in patient care improvement.

Benson C, Schmeling P, Bruins G 1977 A systems approach to evaluation of nursing performance. Nursing Administration Quarterly 1(3): 67–76

Bernhardt J, Schuette L 1975 P.E.T.: A method of evaluating professional nurse performance. Journal of Nursing Administration 5(8): 18–21

Dyer E D, Monson M A, Van Drimmelen J B 1975 What are the relationships of quality patient care to nurses' performance, biographical and personality variables? Psychological Reports 36: 255–266

Haar L P, Hicks J R 1976 Performance appraisal: derivation of effective assessment tools. Journal of Nursing Administration September: 20–29

Hinshaw A S, Oakes D L 1977 Theoretical model-testing: patients', nurses', and physicians' expectations of quality nursing care. In: Communicating Nursing Research 10. Western Interstate Commission for Higher Education. Boulder, Colorado, USA: 163–187

McClure M L 1976 Quality assurance and nursing education: a nursing service director's view. Nursing Outlook (24): 367–369

Quality assurance and peer review 1977 Nursing Administration Quarterly 1: 3

Ramphal M 1974 Peer review. American Journal of Nursing 74: 63–67

Schwirian P M 1978 Evaluating the performance of nurses: a multi-dimensional approach. Nursing Research 27: 347–351

Sheridan J E, Fairchild T J, Kaas M 1983 Assessing the job performance of nursing

home staff. Nursing Research 32(2): 102–107
Smith H L, Mitry N W 1983 Nurses' quality of working life. Nursing Management
    14(1): 14–18
Watson A 1982 Cardiopulmonary resuscitation competencies of nurses. International
    Journal of Nursing Studies 19(2): 99–107

3. Clinical studies

The readings listed below are a sample of the way nurses are using
concepts of quality care in clinical settings. Where the area of
application has not been indicated in the title a notation has been
made in the citation. In the aggregate these papers indicate a movement away from excessive dependence on intuition and tradition
based nursing and increasing efforts to establish a research based
profession. Most of the titles give a fairly good indication of the area
of interest. Some deal with outcome measures (Bowen & Miller;
Cope & Cox; Daubert; Dziurbejko & Larkin; Flaherty & Fitzpatrick; Fortin; Granger; Hendry & Shea; Tanner & Noury; Wilson-
Barnett 1981). Others are process directed (Cope & Cox; Tonkin).
Procedure and technique oriented studies are clearly identified by
titles except the Foster item which deals with volume readings on
plastic urine bags. The Hogstel; Krulik; Pinelli; Shields, Hovey,
Fuller; and Wilson-Barnett (1981) articles all make a direct contribution to nursing knowledge.

Alaszewski A 1978 The situation of nursing administrators in hospitals for the
    mentally handicapped: problems in measuring and evaluating the quality of care.
    Social Science and Medicine 12(2A): 91–97
Autorino G, Hogan M 1977 Application of standards in critical care nursing. In:
    Nicholls M E, Wessells V G (eds) Nursing standards and nursing process.
    Contemporary Publishing Inc, Wakefield, Massachusetts, USA: 125–133
Bergman R 1975 Evaluation of community health nursing. Australian Nurses Journal
    4(Mar): 39–41
Bowen S M, Miller B C 1980 Paternal attachment behavior as related to presence at
    delivery and preparenthood classes: a pilot study. Nursing Research 29: 307–311
Brandt H 1978 Improving the quality of antenatal care. Nursing Mirror 147 August
    10: 7–9
Carl L 1983 Nursing criteria for trauma center site review. Journal of Emergency
    Nursing 9: 74–77
Christoffel T, Loewenthol M 1977 Evaluating the quality of ambulatory health care: a
    review of emerging methods. Medical Care 15: 877–899
Cope D, Cox S 1980 Organization development in a psychiatric hospital: creating
    desirable change. Journal of Advanced Nursing 5: 371–380
Daubert E A 1979 Patient classification system and outcome criteria. Nursing
    Outlook 27: 450–454 (rehabilitation)
Dziurbejko M M, Larkin J C 1978 Including the family in preoperative teaching.
    American Journal of Nursing 78: 1892–1894
Erickson R 1980 Oral temperature differences in relation to thermometer and
    technique. Nursing Research 29: 157–164

Evers H K 1981 Multidisciplinary teams in geriatric wards: myth or reality? Journal of Advanced Nursing 6: 205–214

Final Report of the Joint Committee on Quality Assurance of Ambulatory Health Care for Children and Youth 1975 Criteria for evaluation of ambulatory child health care by chart audit: development and testing of a methodology. Pediatrics 56(supplement): 625–692

Flaherty G G, Fitzpatrick J J 1978 Relaxation technique to increase comfort level of postoperative patients: a preliminary study. Nursing Research 27: 352–355

Flynn B C 1977 Research framework for evaluating community health nursing practice. In: Miller M, Flynn B (eds) Current perspectives in nursing. C V Mosby, St Louis, p 35–45

Flynn B C, Ray D W 1979 Quality assurance in community health nursing. Nursing Outlook 27: 650–653

Forchuk C, Martin M L, Henderson Smith R, Rowe P 1983 Adding a new dimension to accountability. Canadian Nurse 79(6): 42–43 (psychiatry)

Fortin F, Kerouac S 1976 A randomized controlled trial of pre-operative patient education. International Journal of Nursing Studies 13: 11–24

Foster J, Nunnick C 1981 Not to be trusted. Nursing Mirror December 2: 39–40 (neurology)

Given B, Given C W, Simoni L E 1979 Relationships of processes of care to patient outcome. Nursing Research 28(2): 85–93

Govaertz K, Patini E 1981 Attachment behavior of the Egyptian mother. International Journal of Nursing Studies 81(1): 53–60

Granger C V, Greer D S, Liset E, Coulombe J, O'Brien E 1975 Measurement of outcomes of care for stroke patient. Stroke 6(1): 34–41

Grant Higgins P 1982 Measuring nurses' accuracy of estimating blood loss. Journal of Advanced Nursing 7(2): 157–162

Hanson S 1975 Ambulatory nursing standards. Supervisor Nurse December: 10–15

Hardy V M, Capuano E F, Worsman B D 1982 The effect of care programmes on the dependency status of elderly residents in an extended care setting. Journal of Advanced Nursing 7: 295–300

Harrison L 1976 Nursing intervention with the failure-to-thrive family. American Journal of Maternal Child Nursing 1(2): 111–116

Heidt P 1981 Effect of therapeutic touch on anxiety level of hospitalized patients. Nursing Research 30: 32–37

Hendry J M, Shea J A 1980 Pre and postnatal care sought by adolescent mothers. Canadian Journal of Public Health 71: 112–115

Hogstel M O 1979 Use of reality orientation with aging confused patients. Nursing Research 28: 161–165

Krulik T 1980 Successful normalizing tactics of parents of chronically ill children. Journal of Advanced Nursing 5: 573–578

Pinelli J M 1981 A comparison of mothers' concerns regarding the care-taking tasks of newborns with congenital heart disease before and after assuming their care. Journal of Advanced Nursing 6: 261–270

Shields J R, Hovey J K, Fuller S S 1980 A comparison of physostigmine and meperidine in treating emergence excitement. The American Journal of Maternal Child Nursing 5: 170–175

Snyder M 1983 Relation of nursing activities to increases in intracranial pressure. Journal of Advanced Nursing 8(4): 273–279

Speedie G 1983 Nursology of mouth care: preventing, comforting and seeking activities related to mouth care. Journal of Advanced Nursing 8(1): 33–40

Tanner G A, Noury D J 1981 The effect of instruction on control of blood pressure in individuals with essential hypertension. Journal of Advanced Nursing 6: 99–106

Tonkin R S 1979 The REACH Centre: Its history and work III. A study of the quality of paediatric care. Canadian Journal of Public Health 70: 405–414

Warner M, New P K 1976 Towards a health centre evaluation model: workshop report. Health and Welfare Canada, Ottawa
Wilson-Barnett J 1981 Assessment of recovery: with special reference to a study with post-operative cardiac patients. Journal of Advanced Nursing 6: 435–445
Wilson-Barnett J, Carrigy A 1978 Factors influencing patients' emotional reactions to hospitalization. Journal of Advanced Nursing 3: 221–229
Wong J 1979 An exploration of a patient-centered nursing approach in the admission of selected surgical patients: a replicated study. Journal of Advanced Nursing 4: 611–619

4. Improving care in a nursing unit

Nurses have written about their experiences in introducing quality assurance programmes in small head nurse units in which they work. Some worked with management committees (Allison & Kinloch) while others worked as individuals or in informal groupings. Some give a step-by-step description of the process (Gallant & McLane; Laing & Nish), and some describe their contribution as guides (Berg et al; Froebe & Bain).

Allison S, Kinloch K 1981 Four steps to quality assurance. Canadian Nurse 77 (December): 36–40
Aydelotte M K 1975 Quality assurance programs in health care agencies. In: National League for Nursing Quality assessment and patient care 1–11
Berg M, Salisbury I, Josten L, Wilfong M, Tankson E A, Magnuson P, Knoll K 1976 Professionally speaking: starting a system for evaluating quality of care: process and product. American Journal of Maternal Child Nursing 1(3): 141–144
Bergman R, Golander H 1982 Evaluation of care for the aged: a multipurpose guide. Journal of Advanced Nursing 7: 203–210
Chavasse J 1981 From task assignment to patient allocation: a change evaluation. Journal of Advanced Nursing 6(2): 137–145
Dyer E D, Monson M A, Cope M J 1975 Increasing the quality of patientcare through counselling and written goal setting. Nursing Research 24(March–April): 138–144
Froebe D J, Bain R J 1976 Quality assurance programs and controls in nursing. C V Mosby, St Louis, USA
Gallant B, McLane A 1979 Outcome criteria: a process for validation at the unit level. Journal of Nursing Administration (January): 14
Laing M, Nish M 1981 Eight steps to quality assurance. Canadian Nurse 77(November): 22–25

5. Methodology and tools for assessment of care

The articles in this section consist of guides for the development of criteria or standards or specific tools. Examples of tools are included in Atwood, Hinshaw, Carrieri, Lewis & Minckley; Bourbonnais; Haussmann & Hegyvary; Hinshaw & Atwood; D. J. Mason. The Clinton, Denyes, Goodwin, Koto article used Orem's self-care conceptual framework to identify desirable outcomes of care. Two articles present thoughtful analyses of what is going on in this area, how the ideas can be classified or organized for research and applications to nursing care. By Gortner, Bloch, Phillips and Hegyvary & Haussmann (1975) they are highly recommended for those who wish to increase their insight into the development of nursing as a rational and research based profession.

Atwood J R 1980 Developing instruments for measurement of criteria: a research perspective. Nursing Research 29(2): 104–108

Atwood J, Hinshaw A, Carrieri V, Lewis F, Minckley B 1978 Quality of care instruments for indexing important outcomes: examples of tool development. In: Communicating Nursing Research Volume 11 Western Interstate Commission for Higher Education, Boulder, Colorado, USA, p 45–55

Balinsky W, Berger R 1975 A review of research on general health status indexes. Medical Care 13(April): 283–293

Barba M, Bennett B, Shaw W J 1978 The evaluation of patient care through use of ANA's standards of nursing practice. Supervisor Nurse (January): 42 ff

Bourbonnais F 1981 Pain assessment: development of a tool for the nurse and the patient. Journal of Advanced Nursing (6): 277–282

Clinton J F, Denyes M J, Goodwin J O, Koto E M 1977 Developing criterion measures of nursing care: case study of a process. Journal of Nursing Administration (September): 41–45

Cornell S A 1974 Development of an instrument for measuring quality of nursing care. Nursing Research (23): 108–117

Downs F S 1980 Relationship of findings of clinical research and development of criteria. Nursing Research 29(2): 94–97

Ethridge P E, Packard R W 1976 An innovative approach to measurement of quality through utilization of nursing care plans. Journal of Nursing Administration 6(January): 25–31

Gortner S, Bloch D, Phillips T 1976 Contributions of nursing research to patient care. Journal of Advanced Nursing 1: 507–518

Hall B A, Horn B J, Mitsunaga B K, Engle T 1978 Quality of care: selected influences and methodological dilemmas. In: Communicating Nursing Research Volume 11, Western Interstate Commission for Higher Education, Boulder, Colorado USA, p 81–89

Haussmann D, Hegyvary S 1976 Monitoring the quality of nursing care. Australian Nurses Journal 5(February): 29–32, 36

Hegyvary S T 1980 Establishing valid and reliable criteria: an evaluator's perspective. Nursing Research 29(2): 91–93

Hinshaw A, Atwood J 1982 A patient satisfaction instrument: precision by replication. Nursing Research 31(3): 170–175, 191

Horn B J 1980 Establishing valid and reliable criteria: a researcher's perspective. Nursing Research 29(2): 88–90

Horn B J, Swain M A 1978 Criterion measures of nursing care quality. National Center for Health Services Research. US Department of Health, Education and Welfare

Krumme U S 1975 The case for criterion-referenced measurement. Nursing Outlook 23: 764–770

Leatt P, Bay K S, Stinson S M 1981 An instrument for assessing and classifying patients by type of care. Nursing Research 30(3): 145–150

Lindeman C A, Hagan D E 1977 Targeted research: measuring quality of nursing care. In: Communicating Nursing Research Volume 8, Western Interstate

Commission for Higher Education, Boulder, Colorado, USA, p 352–365
Mason D J 1981 An investigation of the influences of selected factors on nurses' inferences of patient suffering. International Journal of Nursing Studies 18(4): 251–259
Mason E J 1978 How to write meaningful nursing standards. Wiley Medical Publication, Wiley and Sons, New York
Runtz S E, Urtel J G 1983 Evaluating your practice via a nursing model. The Nurse Practitioner 8(3): 30ff
Thibodeau J A 1983 Nursing models: analysis and evaluation. Wadsworth Health Sciences Division, Monterey, California

6. Nursing audit

In spite of the difficulty of differentiating nursing outcomes from the outcomes of care from allied health care professionals there is considerable literature on nursing audits. Much of the literature is repetitive but a selection has been listed below. Some authors describe the development of audit programs or development of flow sheets and recording charts. Karch describes the process from developing criteria to using check lists. Miller & Knapp discuss analysis of the data obtained by audit. The Robinson item is an elementary 'how to' book, step-by-step instruction based on the nursing process. Vasey is also elementary, but more concerned with motivation of nurses to participate in audits. Maintenance of good records is covered by several authors including Robinson.

Anderson N 1977 Audit of care processes and patient outcome: one facet of quality assurance. Nursing Administration Quarterly 1(3): 117–128
Apostoles F E, Little M E, Murphy H D 1977 Developing a psychiatric nursing audit. Journal of Psychiatric Nursing and Mental Health Services 15(5): 9–15
Benedikter H 1977 From nursing audit to multidisciplinary audit. National League for Nursing, New York
Dorsey B, Hussa R 1979 Evaluating ambulatory care: three approaches. Journal of Nursing Administration (January): 34–43
Karch A M 1980 Concurrent nursing audit: quality assurance in action. Chas. B. Slack Inc, Thorofare, New Jersey, USA
Laing M 1981 Flow sheets: meeting the charting challenge. Canadian Nurse 77(December): 40–42
Miller M C, Knapp R G 1979 Evaluating quality of care: analytical procedures, monitoring techniques. Aspen Systems Corporation, Germantown, Maryland, USA
Robinson J (ed) 1978 Documenting patient care responsibly. Nursing Skillbook. Intermed Communications Inc, Horsham, Pennsylvania, USA
Trussell P M, Strand N 1978 A comparison of concurrent and retrospective audits on the same patients. Journal of Nursing Administration 8(May): 33–38
Vasey E K 1978 Understanding the nursing audit: how it benefits you. In: Robinson J (ed) Documenting Patient Care Responsibly. Nursing Skillbook. Intermed Communications Inc, Horsham Pennsylvania, USA, p 159–166

Watson A, Mayers M 1976 Evaluating the quality of patient care through
    retrospective chart review. Journal of Nursing Administration 6(March/April):
    17–21
Wilson J 1983 The Canadian hospital accreditation program. Canadian Nurse 79(66):
    48–49

7. Evaluation research

This list is brief but distinctive and important. Luker and Melia provide thoughtful discussions about the *kind* of research in which they have been engaged. The issue of Nursing Research contains 15 papers, 12 by United States nurses who have established reputations as researchers in quality assurance. Three are by non-nurse researchers involved in quality assurance and standards. The papers are issue oriented but many also present methodologies and some instruments. The papers are introduced in an editorial by Norma Lang. The final paper deals with planning for the future in evaluation research.

Luker K 1981 An overview of evaluation research in nursing. Journal of Advanced
    Nursing 6(2): 87–93
Melia K 1982 Tell it as it is — qualitative methodology and nursing research. Journal
    of Advanced Nursing 7(4): 327–335
Nursing Research 1980 Volume 29 (March–April): 68–126 (A special issue devoted to
    evaluation research)

8. Unclassified

Canadian Nurses Association 1980 A definition of nursing practice. Standards for
    nursing practice. Ottawa, Ontario
Chase B A 1982 Nursing service standards as a context for self-assessment. Journal of
    Continuing Education in Nursing 13(4): 26–38
Cox C L, Baker M G 1981 Evaluation: the key to accountability in continuing
    education. Journal of Continuing Education in Nursing 12(1): 11–19
Toward assessment of quality of care in physiotherapy 1979; 1981 Volumes 1 and 2.
    Report of an expert group. Canada Health Services Directorate, Department of
    National Health and Welfare, Ottawa

Index

Accommodation, hospital, 58
Accreditation
  American Nurses Association (ANA),
    75, 76
  Australia, 97
  community sanction and, 32
  quality assurance and, 74, 75, 76
American Nurses Association (ANA)
  accreditation, 75, 76
  certification, 75
  continuing education, 75
  evaluation, partial, 107, 108–109, 117,
    118
  licensure, 75
  peer review, 75, 76
  quality assurance, 69–72, 75–76
  standards, 22–23, 109
Audit, nursing, 197
  outcome criteria and, 75
  Phaneuf, 77, 110–111, 117, 118
  see also Nursing Audit Quality
Australia
  accreditation, 97
  nursing education, 93, 94
  nursing practice, 91–93
Australian Conciliation and Arbitration
  Act, 90

Canadian Hospital Association, staffing
  and, 129–130
Canadian Nurses Association (CNA),
  standards and, 23
Care
  assessment tools, 38, 58–59, 195–196
  discipline-orientated medical model,
    7–8
  emancipatory, 8–10
  holistic approach to patient, 8, 13
  hospitalised adult, 78–79
  medical technology and, 6–7
  models of, 6–10

non-nursing work, 92
plans see Care plans, nursing
segmentation of nursing skills, 10–11
task-orientated, 30-31
team nursing, 11–12, 31
Care plans, nursing
  computer-generated, 175–176
  nurse education and, 101, 102
  nursing process evaluation and, 100
  predeveloped, 60
Change agent role, 9, 14
  nursing research committee and, 49
Clinical research, 43–64, 77–81
  accommodation, hospital, 58
  aged clients in community and, 53
  beliefs of patients, 56–57
  change, implications for, 49–53
  components of care, 57
  continuing education, 60–61
  descriptive, 43–44
  discharge planning, 57
  electronic instrumentation and see
    Electronic instrumentation, data
    collection
  environmental topics,
    57–58
  establishment of research
    environment, 55–56
  expectations of consumers and, 46–47,
    51–52
  experimental, 44
  historical, 43
  impact on environment, 53
  implemenatation, 55–63
  need for, 44–48
  nurses as independent practitioners
    and, 53
  nursing research committee, 48–49
  orientation programs, 61–62
  predeveloped nursing care plans, 60
  providers of care, 44–46
  research process, 151–155

199

research topics, 56–63
resistance to, 53–54
self-assessment of nurses, 58–60
standards of care quality, 58–60
tools, 54–55
users of care, 46–48, 51–52
Community nursing, 3–4
emancipatory approach, 10
nurse operated clinics, 79
primary nursing, 13
quality assurance studies, 79
segmentation of expertise and, 11
team nursing, 12
Computers, 174–188
administrative applications, 178
clinical research and, 52
diabetic monitoring measures and, 172
educational applications, 178
nurse scheduling application *see*
Workload measurement
nursing care plan generation, 175–176
nursing care summary, 176–178
quality assurance and, 103–104, 118
realtime nursing system, 174–175
see also Rush-Medicus
Confidentiality of data, 81–83
Consumer organisations, 7
Continuing education
American Nurses Association (ANA),
75
Australia, 91
clinical research, 60–61
Criteria
outcome *see* Outcome criteria
routine nursing care, 22
standard evaluation, 18–21
Criterion measures (Horn and Swain
method), 110, 114–116, 117, 118

Definitions of nursing, 21–22, 99–104
Delphi index I, 180–182, 185, 186
Delphi index II, 182–183, 186, 187
Diabetic monitoring measures, data
collection and, 168–172
capillary blood glucose monitoring,
168, 171
urine testing, 168, 169
Discharge planning, 57

Education
community sanction and, 31–32
medical model, 101, 102
problem solving approach and, 101
see also Continuing education

Electronic instrumentation, data
collection and, 151–172
calibration, 156–157
compliance to standards, 157
diabetic monitoring measures study,
168–172
noise, instrumental, 156
noise study, 164–167
readout, 156
sensors, 156
sleep study, 158–164
systems isolation, 156
time resolution, 157
Ethics, International Code, 33
Evaluation
care delivered, 26–27
care related to patient's needs, 27–29
criterion measures (Horn and Swain
method), 110, 114–116, 117, 118
definitions, 26
global methods, 110–116
meaning of language and, 24–26
model, 17–18
multidisciplinary approach, 27
multinational programs, 25
nursing practice and, 26–35
partial methods, 107–109, 117
process of nursing, 18, 98, 100
research, 76, 77–78, 198
staffing methods and, 126
structure, 18, 97
tools, 38, 58–59

Family involvement, 9, 13

Gerontological nursing, 81
Goal-directed management, 11

Horn and Swain method, 110, 114–116,
117, 118

Joint Commission for Hospitals
Association, staffing and, 129

Licensure
American Nurses Association (ANA),
75
Australia, 91
community sanction and, 32
standards and, 32
Legislation, community sanction and, 33

Medicus Patient Classification, 135–137
Mental health nursing, 80–81
Méthode d'appréciation de la qualité des soins infirmiers (MAQSI), 110, 116, 117, 118

National League of Nursing, staffing and, 128
National Quality Assurance Committee (NQAC), Australia, 95, 96, 98, 104–105
New Zealand Education and Research Foundation (NERF), standards and, 23
Ninewells index I, 179–180, 185, 186
Ninewells index II, 183–185, 186, 187
Noise level, electronic data collection, 164–167
Nurse operated clinics, 79
Nursing Attention Requirement level (NARvel), 59
Nursing audit *see* Audit, nursing
Nursing Audit Quality, 107–108, 117, 118
Nursing care plan *see* Care plan, nursing
Nursing Performance Review, 76
Nursing Professional Standards Review, 76
Nursing research committee, 48–49

Orientation programs, 61–62
Outcome criteria
  gerontological nursing, 81
  mental health nursing, 80
  nursing audits and, 75
  quality assurance and, 75

Patient classification, 128–130, 132–138, 147
  applications, 141–144
  baseline staffing and, 142
  cost, determination of, 143
  critical indicators of care and, 133–135
  definition, 131–132
  equitable distribution of nursing assignments and, 142–143
  in service education and, 144
  Medicus Patient Classification, 135–137
  monitoring changes in nursing practices and, 143
  nursing task documents and, 133
  profile approach and, 133

Public Health Service Patient Classification, 137–138
  reliability monitoring, 144–145
  selection of process criteria and, 143
  staff orientation and, 144
  staffing and, 128–130
  validity monitoring, 145–146
  variable staffing and, 142
Peer review, 96, 97, 98, 192
  American Nurses Association (ANA), 75, 76
Peer Review Resource Centre (PRRC), Australia, 93
Phaneuf Nursing Audit, 77, 110–111, 117, 118
Primary nursing, 12–15, 35–36
  body of knowledge, 31
  professional authority, 31
Process, nursing
  criteria and, 21
  evaluation and, 26, 100
Profession, nursing as
  characteristics of profession, 30
  clinical research and, 46
  community sanction and, 30, 31–33
  ethical codes and, 30, 33
  primary nursing and, 13, 36
  professional authority and, 30–31
  professional culture and, 30, 35
  research on nurses' role and, 52
  self-regulation of standards and, 22
  systematic body of knowledge and, 30, 31
Professional Development Committee (PDC), Australia, 95, 100, 102, 104
  nursing standards and, 97, 98, 99, 103
Public Health Service Patient Classification, 137–138

Quality assessment, 69
Quality assurance
  accreditation and, 74, 75, 76
  activities of nurses and, 83
  American Nurses Association (ANA), 69–72, 75–76
  appropriate method, 116–118
  Australia, 89–105
  certification, 75
  community health nursing, 79–80
  conceptual framework, 69–72
  confidentiality of data, 81–83
  empirical studies, 77–81
  equipment, 118
  federal influence, 72–73
  gerontological nursing, 81

health promotion for
  mothers/children, 80
hospital-based programs, 4
hospitalized adult, nursing care, 78–79
licensure and, 75
measurement of standards/criteria, 71
mental health nursing, 80–81
National Quality Assurance
  Committee (NQAC), Australia, 95
nurse-midwifery, 80
nursing assignment patterns and,
  78–79
objectives of management and, 117
outcome criteria and, 75
parent-child nursing, 80
peer review and, 75, 76, 96–97
Peer Review Resource Centre
  (PRRC), Australia, 93
personnel investment, 118
political aspects, 52
private sector insurance business and,
  74
Professional Development Committee
  (PDC), Australia, 95
Royal Australian Nursing Federation
  (RANF) program, 96–99
size of institutions and, 117
time investment in programs, 118
time required for corrective action,
  118
time required for starting program,
  117–118
USA, 69–84
voluntary influence, 74–76
Quality patient care scale
  (QUALPACS), 77, 80, 110, 111–112,
  117, 118

Registered Nurses' Association of British
  Columbia (RNABC), 107, 108
Royal Australian Nursing Federation
  (RANF), 90, 93, 94, 95, 104–105
  formulation of nursing standards and,
  97
  quality assurance program, 96–99
Royal College of Nursing (RCN),
  standards and, 23
Rush-Medicus, 78, 110, 112–114, 117,
  118

Selected Attributed Variable Evaluation
  (SAVE), 59
Self assessment, 58–60, 192
Slater Nursing Competency Scale, 77
Sleep, electronic instrumentation study,
  158–164
Staffing, nurse, 123–148
  conceptual framework, 124–127
  historical aspects, 127–130
  levels, 126, 127–128
  methods, 127–146
  nursing care time measurement and,
  138–141
  patient acuity and, 131
  patient assessment and, 131
  patient census and, 129
  patient classification and see Patient
  classification
  patient dependency and, 130–131, 132
  patterns, 126
  professional judgement, allocation of
  resources and, 127, 141–142,
  145–146
  quality of patient care and, 146–147
  requirements for nursing care and,
  131, 132
Standards
  clinical research, 58–60, 103
  computerisation and, 104
  cost estimation and, 103
  criteria and, 18–21
  formulation, 98–99
  as guides for care, 102
  implementation, 99
  national, 21–24
  patient classification system and,
  100–101, 102–103
  process/outcome of care and, 100–104

Task orientated nursing, 30–31
Team nursing, 11–12, 31

Workload measurement
  average care time method, 139–140
  estimation procedures, 140–141
  indices of nursing workload,
  computer-estimated, 178–187
  standard times, 139, 140